Horticulture Science Series – Vol. 1

AROMATIC PLANTS

Horticulture Science Series – Vol. 1

AROMATIC PLANTS

Editor

Prof. K.V. Peter

Professor of Horticulture; Former Vice Chancellor
Kerala Agricultural University, Thrissur - 680 656, Kerala

Authors

Baby P. Skaria (Head)

P.P. Joy, Samuel Mathew, Gracy Mathew, Ancy Joseph (Scientists)

Ragina Joseph (Project Associate)

Kerala Agricultural University
Aromatic and Medicinal Plants Research Station
Odakkali, Asamannoor Post, Ernakulam District, Kerala - 683 549

nipa®

NEW INDIA PUBLISHING AGENCY

New Delhi — 110 034

NEW INDIA PUBLISHING AGENCY
101, Vikas Surya Plaza, CU Block, LSC Market
Pitam Pura, New Delhi – 110 034, India
Email: info@nipabooks.com
Web: www.nipabooks.com

For customer assistance, please contact
Phone: + 91-11-27 34 17 17 Fax: + 91-11- 27 34 16 16
E-Mail: feedbacks@nipabooks.com

ISBN:978-81-89422-45-5

Composed and Designed by NIPA.

National Academy of Agricultural Sciences

Foreword to the Series

Prof. M.S. Swaminathan
President

A second green revolution aimed to meet the food requirements of about 1.8 billion people in India by 2010 is on the anvil. It is, however, imperative that the attainment of the food and nutritional security of this vast population should be combined with the preservation of clean environment. Horticulture encompassing fruits, vegetables, ornamentals, plantation crops, spices, tuber crops, medicinal and aromatic plants, besides harvest and post harvest technologies has a special role in providing food and nutritional security, rural poverty alleviation, employment generation as well as improving the quality of life in our country sides.

India, incidentally, is the world's leading producer of fruits and is followed by Brazil; in vegetables it occupies the second place after China. India is also the land of spices producing an array of such crops. Indeed, the varied eco-climatic conditions prevailing in different parts of this country make it possible to grow a disparate range of fruits, vegetables and ornamental plants at any time of the year.

NASC Complex
DPS Marg, P.O. Pusa
New Delhi - 110 012

Tel 91-11-25846055
91-11-25846051
91-11-25846052
91-11-25846053
Fax 91-11-25846054
E-mail naas@vsnl.com
Website
//education.vsnl.com/naas
P.O. Box No. 11325

With the opening up of the world markets under the WTO regime, however, international competition for Indian horticultural products has reached unprecedented levels, necessitating strategies for quality control, branding and product development. In particular, production of organic vegetables, fruits, cut flowers, pepper, ginger, tea, coffee, rubber and cardamom has of late become demand-driven.

Among the inputs for higher productivity in horticultural crops, seeds and quality planting materials account for 20 to 30% of the potential productivity. Transgenic crops are a reality now; yet the technology for production of transgenics is a far cry in many situations. Furthermore, water is a limiting factor in several environments. Hence, water saving crops and those with high water use efficiency should be preferred over other crops and management practices. Likewise, the use of biodegradable plastics, biological control of pests and diseases, and new product development are receiving increasing attention, especially in the arena of horticultural crop management. Marketing, village knowledge centres, packaging industry, zero-energy cool chambers and a variety of economically viable technologies, however, remain to be streamlined.

The present series on **Horticulture Science**, edited by Professor K.V. Peter, Vice-Chancellor, Kerala Agricultural University, is a good compilation of the scientific knowledge and experience gained over the past 30 years in this important sector of Indian economy. Each volume in the series is authored by accomplished researchers such as Dr. T. Radha and Dr. Lyla Mathew (*Fruit Crops*), Dr. T.R. Gopalakrishnan (*Vegetable Crops*), Dr. P.K. Valsalakumari (*Flowering Trees*), Dr. P.K. Rajeevan (*Ornamental Flowers*), Dr. Alice Kurian and Dr. K.V. Peter (*Plantation Crops*), Dr. Alice Kurian and Dr. M. Asha Sankar (*Medicinal Plants*), Dr. Baby P. Skaria (*Aromatic Plants*), Dr. P. Palaniswamy (*Tuber Crops*), Dr. E.V. Nybe and Dr. N. Mini Raj and Dr. K.V. Peter (*Spices*), Dr. S. Rajan and Dr. Baby Lizzy Markose (*Propagation of Horticultural Crops*), Dr. M.N. Sheela (*Flower for Trade*), Dr. Kanthaswamy (*Coastal Horticulture*), Dr. K.P. Sudheer and Dr. V. Indira (*Post Harvest Technology of Horticultural Crops*), Dr. K. Pradeep, N. Kumar and Dr. T.R. Gopalakrishnan (*Management of Horticultural Crops*) and Professor K.V. Peter (*Basics of Horticulture*). The New India Publishing Agency, New Delhi deserves special credit for publishing this useful series, especially in view of the fact that the Government of India is currently implementing the National Horticultural Mission.

Prof. M.S. Swaminathan, FNA, FRS
President, National Academy of Agricultural Sciences
and Chairman, National Commission on Farmers
New Delhi 110 012

Dr. S. Natesh
Adviser Gr. I
(Scientist-H)

Telefax : 2436 4064
E-mail : natesh@dbt.nic.in

भारत सरकार
विज्ञान और प्रौद्योगिकी मंत्रालय बायोटेक्नोलॉजी विभाग
ब्लाक–2, 7 वां तल, सी0 जी0. ओ0 कम्पलेक्स
लोदी रोड, नई दिल्ली-110003

GOVERNMENT OF INDIA
MINISTRY OF SCIENCE & TECHNOLOGY
DEPARTMENT OF BIOTECHNOLOGY
Block-2, 7th Floor C.G.O. Complex
Lodi Road, New Delhi-110 003

Foreword

The art and science of perfumery and flavouring in India have flourished since ancient times with the advent of civilization and perfected under the patronage of Moghul Emperors. Perfumers of Kannauj practised distillation from sandalwood, agarwood, and rose and developed finest scents and attars of the time.

Aromatic plants contain odorous volatile substances, which occur as essential oil, green exudate, balsam and oleoresin in one or more parts, namely, root, wood, stem, foliage, flowers and fruit. Essential oils are highly concentrated secondary metabolites of diverse functions in plant system. They constitute hundreds of organic compounds including terpenoids, benzenoids, organic sulphur and introgenuous compounds, which work at different levels.

Aromatic herbs, oils and aroma chemicals are extensively used in perfumery, flavouring, cosmetic and drug industries. Moreover, they have now become indispensable ingredients of every human activity. Eco-tourism coupled with ayurvedic rejuvenation and aromatherapy is increasingly attracting tourists from world over to India. India is famous for superior quality essential oils, spice oils and oleoresins. Cochin oil is the finest lemongrass oil in the world market.

India is probably the only country, which can grow all the major essential oil yielding plants. India's bio-diversity is unmatched with the presence of 16 different agro climatic zones, 10 vegetation zones, 25 biotic provinces and 426 biomes. Deforestation and indiscriminate collection are dwindling natural resources. Unless large-scale scientific cultivation is taken up to augment supply of raw materials, the flavour, fragrance, cosmetic and pharmaceutical industries will be in doldrums in a near future. There is great scope for commercial cultivation of these crops owing to immense demand for newer and specific aroma chemicals for development of novel and more stimulating flavours and fragrances.

Cheaper synthetics may overtake naturals at times, but the chemical complexity of each essential oil created in nature can no way be matched. However, greater sophistication of man-made fragrances based on synthetic molecules, high tech blending and quality control, the task has become more challenging.

The present book 'Aromatic Plants' is a well conceived and well thought out publication covering the whole gamut from cultivation to utilization of aromatic plants. This is a comprehensive publication comprising chapters on history, importance and scope of aromatic plants, their classification, extraction, utilization, quality assurance and aromatherapy. Major and minor aromatic crops are dealt with separately. The work is an excellent practical manual and reference book for students, researchers, entrepreneurs, extension personnel and farmers. I hope it will inspire more and more people to strengthen the flavour, fragrance, cosmetic and sector and safeguard our ancientn heritage and wisdom.

Dr. Baby P. Skaria and his colleagues at the Aromatic and Medicinal Plants Research Centre, Kerala Agricultural University, Odakkali deserve to be congratulated for bringing out this timely publication. I would also like to compliment Professor K.V. Peter for editing and co-ordinating the book. I am sure this excellent book will be well accepted by all concerned.

S. Natesh

Preface

The world of fragrance is highly varied and fascinating. Of about 18000 plant species in India, about 1500 are aromatic, producing essential oil in one or another plant organ, may be root, leaves, flowers and some times bark or wood. In some cases, whole plant is aromatic. They are so diverse in their aroma owing to characteristic volatile oil constituents. It is practically impossible to precisely describe an aroma. Predominance of one or a few volatile constituents such as citral, citronellal and geraniol makes them so unique by imparting pleasing note to them. More than a hundred aromatic oils are in global trade. Essential oils and aroma chemicals derived from them are widely used in cosmetics and food industries. Most of them possess medicinal properties and are used in balms, ointments and skin care products. Their use in health and beauty care can be traced back to 1500 years before Christ. Incenses were used by ancient Egyptians as offering to God in sacred rites. Aroma chemicals were used in auspicious occasions and were presented as precious gifts to kings and emperors. Myrrh and frankincense were among the offering to infant Jesus by the Kings from the East. For Arabs, who developed distillation techniques, fragrances were an integral part of daily life and personal well being. In history, it was always associated with quality of life and expression of reverence. The essence obtained from aromatic crops is no longer an article of luxury but an essential requisite for personal and social hygiene and health care. This hand book on aromatic plants is intended to introduce this wonderful category of crops to undergraduates and to create awareness and interest among progressive farmers to undertake scientific cultivation in order to meet the growing demand for essential oils and natural aroma chemicals.

The book is designed as a concise reference for students and a section of entrepreneurs and farmers. The introductory chapter deals with the importance of aromatic crops and their close association with human health and beauty care from time immemorial. History of development of cultivation and aroma based industries in different regions of the world is described to emphasize their significance, scope and role in increasing the quality of human life. Classification of aromatic crops based on their climatic requirement, growth habit and floral morphology elaborated in succeeding chapter will be of great interest to students as well as farmers engaged in their cultivation.

Traditional as well as modern techniques employed for efficient extraction of volatile oils from different plant materials are given in Chapter III. Quality of oil is found to vary significantly with ecotypes, season, time of collection, crop maturity and weather conditions prevailing during the growth period, extraction method and duration of extraction process. Conditions and duration of storage also have a bearing on quality of essential oil. This necessitates development and imposition of appropriate quality standards in trade. These aspects are discussed in Chapter IV.

Aromatic oils and their derivatives and combinations occupy a covetable position in holistic medicines such as aromatherapy. A separate chapter is assigned to furnish information on its wide spread application in aromatherapy to relieve stress and to rejuvenate body.

The number of aromatic crops in commercial cultivation is limited and information available on cultivation technique is scanty especially in case of lesser known aromatic crops. Based on priority use, aromatic plants are grouped as major and other sources of aromatic oils and arranged alphabetically in chapters VI and VII, respectively. Distribution, botany, types and varieties, soil and climatic requirements, cultivation, common pests and diseases and their management, extraction and utilization of essential oil and their physico-chemical properties are covered for each. At the end of each chapter, suggested readings are appended to benefit those interested in learning further about the aspects dealt with.

Information from varying sources was gathered to provide updated literature and the book is presented as a compendium of information on the treasure house of aromatic plants. We hope that the book will serve as a concise and authentic source of basic information on all aspects of essential oil crops cultivation and utilization. With the expectation that this small effort on aromatic plants will pave way for boosting the enchanting world of natural aroma, this book is presented for the readers.

Authors

Acknowledgement

We gratefully acknowledge the fraternity of scientists, researchers, industrialists, entrepreneurs and farmers who have toiled hard and spent immense resources over years in generating and verifying valuable knowledge on aromatic crops which formed foundation for this publication. We are especially thankful to the scientists of Kerala Agricultural University, Thrissur; Kerala Forest Research Institute, Thrissur, Tropical Botanical Garden & Research Institute, Thiruvananthapuram, Central Institute of Medicinal & Aromatic Plants, Lucknow, Regional Research Laboratories and Research and Development Units of Industries for sharing their vast experience and knowledge resource in the field to build a strong backbone to this book.

We express our deepest sense of gratitude to Dr. K.V. Peter, Vice-Chancellor, Kerala Agricultural University and Editor of this book for his persistent inspiration, expert guidance and wholehearted support throughout the preparation of the book, besides the huge editorial task to make it a unique publication.

We profusely thank Dr. D. Alexander, Director of Research and Dr. M.K. Sheela, Director of Extension, Kerala Agricultural University for their relentless encouragement and valuable guidance in bringing out this matchless publication.

Dr. M. Tamil Selven, Director, Directorate of Arecanut and Spices Development, Calicut, Ministry of Agriculture, Government of India who has always been our morale booster without whose keen interest, valuable advice and timely help this would not have been a success. We express our deepest gratitude to him.

We are highly grateful to Dr. S. Natesh, Adviser, Gr. I, Scientist (H), Govt. of India, Ministry of Science & Technology, Department of Biotechnology, Lodi Road, New Delhi-110 003 for gracing us with a thoughtful foreword for this book.

Staff and labourers of Aromatic and Medicinal Plants Research Station, Odakkali, Kerala have always been with us in one accord with untiring commitment and positive criticisms for development of aromatic fraternity of the country. We record in golden letters our indebtedness to them.

We also profusely thank M/s. New India Publishing Agency, New Delhi for designing and printing the book most lucidly.

We sincerely thank each and every one who participated in this endeavour to make this publication an excellent practical manual and reference book in the field.

Authors

Contents

1

History, Importance and Scope of Aromatic Plants

Aromatic plants synthesise and preserve a variety of biochemical products, many of which are extractable and useful as chemical feed stocks or as raw materials for various scientific investigations. Many secondary metabolites of plants are commercially important and find use in a number of perfumery, flavouring and pharmaceutical compounds. The characteristic property of the plant is due to a variety of complex chemical compounds and hence aromatic plants are generally referred to as 'natural bio-chemical factories' or 'chemical goldmines'. Not all these natural chemicals can be synthesised in laboratory. Aromatic plants possess odorous volatile substances, which occur as essential oil, green exudate, balsam and oleoresin in one or more parts, namely, root, wood, stem, foliage, flower and fruit. The term essential oil is concomitant to fragrance or perfumes because these fragrances are oily in nature and they represent the essence or active constituents of plants. They are called volatile or ethereal oils as they evaporate when exposed to air at ordinary temperatures.

History

Perfumes were used by the Egyptians as an offering to their Gods. Their temples had a small room in which all the aromatic products were prepared and kept. The 'Hebrew' knowledge of aromatic perfumes was gleaned from their countrymen who were prisoners of the Egyptians. They used incense and aromatic oils for an ointment in religious ceremonies and to ward off evil spirits. They also knew about therapeutic properties and applied these in medical care. Hippocrates attempted to stop spread of plague in Athens with aromatic fumigation in streets.

Advent of Christianity and fall of Roman Empire saw a decline in use of perfumes and aromatics. Later on, Arabs began to use essence and they perfected the art of distillation. In 10th century AD, a well-known Arab scholar,

Avienna was credited with first distillation though it had, in fact already been done before his time. Essence of rose was distilled along with that of many other plants and these extracts were widely used as remedies for various ailments. India enjoyed a pre-eminent position in the manufacture of superior perfumes and aromatics since ancient times and the industry flourished and grew considerably. Cheaper synthetic substitutes took over naturals due to its price advantage but it is now realized that the complex chemical components of aromatic oils created in nature can no way be matched. World of essential oils has now come out from yesterday's narrow definition of status symbol of luxury to indispensable necessities of common man in every day life, from the cradle of infancy to silence of grave. Aromatic temptations have even changed course of world history and civilisations. Herbal 'renaissance' is happening globally and herbal products today symbolise safety in contrast to synthetics. Over three-quarters of world population relies mainly on plants and plant extracts for health care. More than 30% of entire plant species, at one time or other, were used for medicinal purposes. Chinese, Indian, Arabian and other traditional systems of medicines make extensive use of about 5000 plants.

Revenue potential

Demand and price of herbal products and essential oils are increasing consistently in national and international markets due to strong pro-consumer movement. In the world of fragrance and flavour industry, essential oils contribute to about 17%. Extent of usage of essential oils is 55-60% for flavours in food industry, 15-21% for fragrances in perfumery/cosmetic industry, 10-20% as starting material for isolation of components, 5-10% as active substances in pharmaceutical preparations and 2-5% for natural products. USA, France and Germany are in the forefront of essential oil trade. It is estimated that annual turnover of perfumery, cosmetic and flavour industry exceeds US $ 6 billion comprising more than 100 essential oils, the world over. In India, estimated production of perfumery raw material is around 5000 t/annum valued at Rs. 400 crores. Annual earning of foreign exchange from their export is around Rs. 130 crores. About 90% of India's requirement of essential oils is met from indigenous production and remaining from import. India ranks 28th in imports and 14th in exports in global trade of essential oils. India's contribution is only 1.1% in exports and 0.7% in import. Home demand for essential oils is projected at 15000 tonnes and export targeted at 3400 tonnes. Production of spice oils and oleoresins during 1998-99 was 2625 tonnes valued at Rs. 293 crores. Annual production of medicinal and aromatic plant's raw material is worth about Rs.200 crores. World production is likely to touch US $5 trillion by 2050.

Present pattern of production and trade of essential oils and aroma chemicals is characterized by factors such as fluctuations in demand and price, competitiveness and instability in supplies, with progressive increase in number of producers in many developing and industrialized countries. China, Brazil and Indonesia are strong competitors for Indian aromatic oils. There is significant expansion in range of essential oil usage and their products during last decade.

Of 18,000 native species found in India, 1500 species contain aromas. Only 65 of them have large and consistent demand in world trade and are hence cultivated. Estimated annual area under aromatic crops in India is more than 20,000 hectares with a production of 1,500 tonnes. Average productivity of essential oil crops is 75kg/ha. Major plant sources of essential oils and aroma chemicals are given below.

Major commercial sources of essential oils, oleoresins and aroma chemicals

Item	Botanical source	Part used	Chemical constituents	Uses
Ajowan	*Trachyspermum ammi*	Seed	Thymol	Flavouring foods, soft drinks, confectionery, medicinal
American Basil	*Ocimum americanum*	Herb	Methyl chavicol, citral, linalool	Pharmaceuticals, aroma chemicals, ointments, balms
Anise	*Pimpinella anisum*	Fruit	Anethol, Methyl chavicol	Flavouring soft drinks, confectionery, pharmaceuticals
Bergamot	*Citrus bergamia*	Fruit	Linalyl acetate, linalool, limonene	Citrus soft drinks flavour, 'East Gray' tea flavour
Camphor	*Cinnamomum camphora*	Wood, leaves	Safrole, piperitone, sabinene, eugenol	Pharmaceuticals, incense, balms
Basil	*Ocimum basilicum*	Leaf twig	Camphor	Pharmaceuticals, aroma chemicals, ointments, balms
Perennial chilli	*Capsicum frutescens*	Fruit	Capsicin	Seasoning, blending
Caraway	*Carum carvi*	Seed	Carvone, limonene	Flavouring
Cardamom	*Elettaria cardamomum*	Seed	Cineole, terpineol, limonene, cymene	Flavouring soft drinks, confectionery
Cassia	*Cinnamomum cassia*	Bark, leaf	Eugenol, iso-eugenol, methyl eugenol	Flavouring

Contd...

Table Contd...

Cedar wood	*Cedrus deodara*	Wood, saw dust, root	Cedrol, cedryl acetate	In perfumery for fixative effects and unique odour, soaps
Celery	*Apium graveolens*	Herb	Limonene	Flavouring foods, pharmaceuticals
Chamomile	*Matricaria chamomilla*	Herb	Chamazulene, azulene, farnesene	Flavouring foods, medicinal
Champak	*Michelia champaka*	Flowers	Cineole, iso-eugenol	Perfumery, cosmetics
Cinnamon	*Cinnamomum verum*	Bark, leaf	Eugenol, iso-eugenol, methyl eugenol	Seasoning, blending, natural flavours, pharmaceuticals
Citronella Ceylon	*Cymbopogon nardus*	Grass	Citronellal, geraniol, citronellol, geranyl acetate	Perfumery chemicals, soap, cosmetics, flavouring
Citronella Java	*Cymbopogon winterianus*	Grass	Citronellal, citronellol, geraniol, geranyl acetate	Perfumery chemicals, soap, cosmetics, flavouring
Clarysage	*Salvia sclarea*	herb	Linalool, ocimene, nerol, geraniol	Flavouring soft drinks and liquors
Clocimum	*Ocimum gratissimum*	Leaf twig	Eugenol, methyl chavicol, methyl cinnamate, linalool	Pharmaceuticals, aroma chemicals, ointments, balms
Clove	*Eugenia caryophyllus*	Bud, leaf, Stem	Eugenol, caryophyllene, humulene	Seasoning, blending, natural flavours, pharmaceuticals
Coriander	*Coriandrum sativum*	Herb, Seed	Linalool, α-pinene, phellandrene, camphor	Flavouring of alcoholic drinks (eg. Gin), meat seasoning, curry blends
Cumin	*Cuminum cyminum*	Seed	Cuminyl alcohol, cuminaldehyde	Seasoning curries, natural flavours
Davana	*Artemisia pallens*	Flowering tops	Davanone, fenchyl alcohol, davanofuran	Natural flavours, beverages, high grade perfumery
Dill	*Anethum sowa*	Herb, seed	Dillapiol	Seasoning pickles, aroma chemicals, pharmaceuticals
Eucalyptus	*Eucalyptus citriodora*	Leaf twig	Citronellal, citronellol, cineole, iso-pulegol	Disinfectants, germicides, soap, cosmetics
Eucalyptus	*Eucalyptus globulus*	Leaf twig	Cineole, caryophyllene, camphene, sabinene, myrcene	Blending, medicinal

Contd...

Table Contd...

Fennel	*Foeniculum vulgare*	Fruit	Anethole, fenchone,	Seasoning, blending, natural flavours, pharmaceuticals
French Basil	*Ocimum basilicum*	Herb	Methyl chavicol, methyl cinnamate, eugenol, linalool	Pharmaceuticals, aroma chemicals, ointments, balms
Geranium	*Pelargonium graveolens*	Leaf twig	Geraniol, citronellol, linalool, iso-menthone	In all kinds of scents
Ginger	*Zingiber officinale*	Rhizome	Zingiberene, zingerone, arcurcumene, farnesene	Seasoning, blending, natural flavours, pharmaceuticals
Hoary Basil	*Ocimum canum*	Herb	Linalool, camphor	Pharmaceuticals, aroma chemicals, ointments, balms
Holy/ sacred Basil	*Ocimum tenuiflorum (O. sanctum)*	Leaf twig	Eugenol, methyl chavicol, methyl cinnamate, linalool	Pharmaceuticals, aroma chemicals, ointments, balms
Japanese mint	*Mentha arvensis*	Leaf twig	Menthol, menthone, terpenes	Flavouring tooth pastes, candies, ointments, tobacco, cough syrups
Jasmine	*Jasminum officinale*	Flower	Benzyl acetate, linalool, linayl acetate, jasmone	Natural flavours
Galangal	*Kaempferia galanga*	Rhizome	Ethyl-trans-p-methoxy cinnamate, pentadecane, cineole, carene, borneol	Medicinal, preservation
Lavender	*Lavandula officinalis*	Flower	Linalool, linalyl acetate	Perfumery, soap, antiseptic, insecticides
Lemongrass	*Cymbopogon flexuosus*	Grass	Citral, linalool, geraniol	Lemon flavour, seasoning
Linaloe	*Bursera delpechiana*	Wood, berries, leaves	Linalool, linalyl acetate, methyl heptanol	Aroma chemicals, perfumery
Marigolds	*Tagetes erecta*	Flowers	Tagetone, linalool, limonene, linalyl acetate	Fragrance, cosmetics
Nutmeg	*Myristica fragrans*	Fruit, seed, aril, leaf	Trimyristin, pinene, camphene, myristicin	Cola flavour, meat seasoning, baking, natural flavours
Ocimum	*Ocimum viride*	Leaf twig	Thymol	Pharmaceuticals, aroma chemicals, ointments, balms

Contd...

Table Contd...

Orange sweet	*Citrus sinensis*	Fruit	Citral	Lemon flavour, seasoning
Palmarosa	*Cymbopogon martinii*	Flowering tops	Geraniol, geranyl acetate, citronellol, linalool	Flavouring tobacco, soap, high grade perfumery
Paprika	*Capsicum annum*	Fruit	Capsanthin	Seasoning, blending, natural flavours
Patchouli	*Pogostemon patchouli*	Leaf twig	Patchoulinol, caryophyllene	Flavouring non-alcoholic beverages, perfumes, soaps, cosmetics
Pepper	*Piper nigrum*	Seed	Piperine	Seasoning, blending, natural flavours, pharmaceuticals
Peppermint	*Mentha piperita*	Twig	Menthol	Flavouring, medicinal
Pimenta/ Allspice	*Pimenta dioica*	Leaf, fruit	Eugenol, α-pinene	Seasoning, blending, natural flavours
Rose	*Rosa damascena*	Flower	Citronellol, geraniol, nerol, linalool	Perfumery, cosmetics, flavouring soft drinks, pharmaceuticals
Rosemary	*Rosmarinus officinalis*	Leaves	Pinene, cineole, linalool	Seasoning blends, medicinal
Sandalwood	*Santalum album*	Heart wood	Santalol, santalene, curcumene, farnesene	Perfumery, soaps, detergents, shampoo
Spearmint	*Mentha spicata*	Herb	Menthol	Chewing gum, oral hygiene
Thyme	*Thymus vulgaris*	Fruit	Thymol, cymene, linalool, limonene, cineole	Aromatherapy, flavouring tooth pastes, candies, ointments, cough syrups
Tuberose	*Polyanthes tuberosa*	Flowers	Geraniol, nerol, farnesol	Perfumery, cosmetics
Turmeric	*Curcuma longa*	Rhizome	Curcumin	Seasoning, blending, natural flavours, pharmaceuticals
Vetiver	*Vetiveria zizanioides*	Roots	Vetiverol, vetivone, eudesmol	Perfumery, medicinal
Ylang-ylang	*Cananga odorata*	Flowers	Cineole	Perfumery, soaps, detergents, shampoo

Pepper, ginger, cardamom, saffron, clove, fenugreek, cumin seed and celery seed oils and oleoresins are primarily used for flavouring. The most traded essential oils are mints, basil, orange, clove leaf, citronella, lemongrass, sandal wood, eucalyptus, geranium, lavender, jasmine and tuberose. Essential oils of ginger, sandalwood, lemongrass, jasmine and tuberose are exported from India to Russia, USA, France, Germany, Britain, Netherlands, Australia and Gulf countries, though 'traditional oils' like sandalwood and lemongrass are showing a downward trend. Domestic requirements of oils of basil, sandalwood, cumin seed, dill seed and juniper are fully met from indigenous production. However, lavender, patchouli, clove, nutmeg, geranium and rose oils are still imported from China, Brazil, Turkey, Bulgaria, Australia, Indonesia and Sri Lanka for meeting the industrial requirement.

Industrial significance

Essential oils and aroma chemicals constitute a major group of industrial products. These oils form indispensable ingredients of necessities in many spheres of human activity. They are adjuncts of cosmetics, soaps, pharmaceuticals, perfumery, confectionery, ice-creams, aerated waters, disinfectants, tobacco, agarbathis and a host of related products. However, with recent advances in organic chemistry, synthetics have outnumbered naturals in a ratio of about 200:1 due to limitations in availability of the latter in sufficient quantities at a steady price over a period of time. Naturals are seen as good and safe. Concern for nature and love for all things which are basic and natural is spearheading to a green movement of everything natural and nature-based consumer products all over the world. A future shock awaits the industry as the already dwindling world resources of coal and petroleum on one side and the philosophy of going back to nature is gaining acceptance internationally on the other side. This reverts dependence of industry on petrochemicals to perennial source of naturals. Thus, because of a large spectrum of usage in the everyday life of man, essential oil and aroma industry has a bright future.

In today's world of consumer boom, role of essential oils increased many folds. Apart from hitherto known applications of essential oils, more and more areas are opening up which will benefit the industry. Use of essential oils in therapeutics is becoming popular in Japan and European countries. Aromatherapy involves use of essential oils and aromatics derived from plants to cure diseases. A few of the essential oils are in many ways better than antibiotics due to their safety and wide spectrum of activity. Synergistic activity of essential oils needs further probe. Application of essential oils in agriculture as antifeedants, repellents, botanical insecticides, natural herbicides and growth boosters are still open to fascinating realms of research. Production of secondary metabolites in bioreactors under controlled conditions using cell and tissue culture offers exciting frontiers of future research.

Medicinal uses

Aromatic herbs, oils and aroma chemicals are not only utilized extensively in perfumery, flavouring and cosmetic industries or as starting material for synthesis of other compounds, but some of them have remarkable application as therapeutic agents in pharmaceutical and drug industries. Medicinal properties of aromatic herbs and spices are used in traditional medicine from time immemorial. Black pepper, ginger, turmeric, clove, cinnamon, galangal, sweet flag, cumin, coriander, sandal wood and vetiver find wide use in ayurvedic medicines. For therapeutic purpose, they are administered as inhalation, oral intake, massages, gargles and mouth wash. Essential oils find varied uses as carminative, antiseptic, sedatives, CNS stimulants, adaptogens, bronchodilators, antistress and muscle relaxants. Many aromatic herbs, essential oils and aroma chemicals are mentioned as official drugs in various pharmacopoeias, pharmaceutical codices and formularies of different countries. Various pharmaceutical aspects, quality, formulations, medicinal/ therapeutic uses and adverse and toxicological effects of these items are dealt with in different editions of such official and authoritative publications.

Essential oils and aroma chemicals stimulate sense of smell and emotional centre of brain. A rhythm exhibiting calmness is produced when an essential oil having sedative effect is inhaled. A few of the important essential oils that produce sedative effects are sandalwood, lavender, bergamot, mint, chamomile and sweet marjoram. Essential oils, which possess sedative action, help in overcoming emotional and mental stresses. Oils of cedar wood and lemon possess antistress property. Essential oils like basil, clove, jasmine, peppermint, ylang-ylang, achillea and cajuput show opposite activity and stimulate central nervous system (CNS) causing alertness and inducing a feeling of well being. Adaptogenic properties are reported for oil of geranium and ylang ylang. Oils of eucalyptus, cinnamon and peppermint are very good for congestive respiratory disorders, while ginger, black pepper, pumilio pine, angelica and eucalyptus oils provide relief in bronchitis and cough. Although a majority of essential oils exhibit antimicrobial properties, some essential oils are exceptionally good and are commonly used as antiseptics. Essential oils stimulate healing process in body when they are applied topically both in local and all body massage. They increase blood flow in skin. Oil of allspice acts as excellent muscle relaxant when applied topically. Carminative properties are observed in many essential oils, particularly in those obtained from fruits of family Apiaceae. Besides, oils of spear mint, peppermint and chamomile to have carminative properties. Oils of clove and thyme provide relief from flatulent colic, since they have antispasmodic property. Clove oils possess analgesic activity and are used extensively in toothache. Leaves of ocimum species are used for a variety of medicinal uses like stimulant and neurotonic, decongestant, uterus and prostate ailments, antiatherosclerosis, viral encephalitis, polio-mellitus, tropical viral infections, varicose veins and circulatory problems. Active constituent in the essential oil of this plant species are methyl chavicol, linalool, eugenol, methyl cinnamate, etc.

Recent pharmaceutical studies showed that oil of achillea and nutmeg causes haemostatic and prostaglandin inhibition activity, respectively. Anti carcinogenic property has been reported for essential oils of cumin and basil and these can be used as protective agents against carcinogenesis. Also methanol extracts of allspice, marjoram, tarragon and thyme strongly inhibited platelet aggregation induced by collagen in humans. Properties of essential oils usually depend upon their major constituents. A majority of them, especially alcoholic compounds, show antimicrobial property. Chamazuline, a terpene isolated from chamomile oil, besides having antimicrobial activity, also exhibits anti-inflammatory activities. Most of esters and aldehydes possess both antimicrobial and sedative activities. Ketones like camphor, pinocarpene, carvone and menthone are very useful in respiratory disorders. Cineole, a peroxide compound isolated mainly from eucalyptus oils has wide range of activities. They have antimicrobial and expectorant properties and induce secretion of live enzymes. Anethole, ether, mainly isolated from anise oil is a good carminative and expectorant. Another ether compound, apiole isolated from parsley has emmenagogue property.

The structurally related terpinoides; citral, citronellol, geranyl acetate, linalool and linalyl acetate are rapidly absorbed in body. At low doses, decarboxylisation is the major metabolic route; at high doses, some terpinoides may be extracted unchanged. An estimated acceptable daily intake of up to 500 μg/kg body weight is established for citral, geranyl acetate, citronellol, linalool, linalyl acetate expressed as citral.

Extraction

Essential oils are obtained by distillation, usually with water or steam or as in case of citrus fruits, by a mechanical process. *Concretes* are odorous concentrates obtained from fresh plant materials of low resinous content by extraction with a volatile non-aqueous solvent, followed by removal of solvent by evaporation at moderate temperatures and under partial vacuum. Concretes are usually waxy solids. *Absolutes* are highly concentrated perfumery materials obtained from concretes by repeated extraction with ethyl alcohol followed by chilling of extract (to precipitate waxes and non odorous matter), filtration or centrifugation of remaining alcohol solution and finally removal of most of alcohol by evaporation at moderate temperatures and under partial vacuum. Absolutes are usually liquids and are entirely soluble in alcohol. *Spice oleoresins* are obtained from dried spices by extraction with a volatile non-aqueous solvent, followed by removal of solvent by evaporation under partial vacuum. Oleoresins contain aroma and flavour of spice (including any non-volatile principles, unlike spice essential oils) in a concentrated form and are usually viscous liquids or semi-solid materials. They should be distinguished from *spice aquaresins*, which have closely related applications but which are extracted with aqueous alcohol rather than with volatile solvents.

Chemical nature

Volatile oils are complex mixtures varying widely in their composition. The characteristic odour and flavour are mainly due to oxygenated organic compounds. Only a few possess a single component in very high percentage, *viz,* santalol in sandalwood oil, citral in lemongrass oil, geraniol in palmarosa oil and eugenol in cinnamon leaf oil. Essential oils largely comprise following major compounds.

I. *Terpenoids*

1. *Monoterpenoids:* Pinenes, ocimenes, limonene, citral, geraniol, linalool, camphor, menthol
2. *Sesquiterpenoids:* Caryophyllene, germacrene, cadinene, longifolene, thujopsene, aromadendrene, zizaene, cedrol, farnesol, ishwarone, santalol, cycloscychellene
3. *Diterpenoids:* Abietic, podocarpic and labdanolic acids; abietol, pimarinol, devadarool, ginkolide, sciarcol, manool
4. *Sesterterpenoids:* Geranyl farnesol, geranyl nerolidol
5. *Triterpenoids:* Squalene, panaxatriol, odoratol, jasminol
6. *Tetraterpenoids:* Carotenes, lycopene

II. *Benzenoids:* Benzaldehyde, benzyl-alcohol, acetate, benzoate, salicylate, methyl-cinnamate, chavicol, eugenol, isoeugenol, cinnamaldehyde, cinnamyl alcohol, cinnamyl cinnamate, eugenol, isoeugenol, eugenyl acetate, isoeugenyl acetate, vanillin, anisaldehyde, anethole, asarone, elemicin, dillapiole, apiole, safrole, chavibetol

III. *Organic sulphur compounds:* Disulphides, sulphoxides

IV. *Nitrogenous compounds:* gueriol pyridine, dehydroguaiol pyridine, guaipyridine, epiguaipyridine, methyl-N-methyl anthranilate, indole, pyrazines, pyridines.

Cultivation and utilisation

Cultivation and processing of essential oil bearing and medicinal plants form an important area in international agri-business with an estimated annual growth rate of 7-10%. India enjoys variety of climatic conditions suitable for growth of a large number of such plants. Availability of huge scientifically trained manpower, improved technologies, plant varieties, processing technologies and relatively cheaper cost of production now place India in a condition favourable for it to become a major producer, processor and supplier of aromatic and medicinal plant materials in the world market. To further augment the quality end product and maintain the acquired status, several steps will have to be taken which include up-gradation of the post harvest technologies and quality control measures, steering of the ongoing innovative R&D programmes and strengthening of liaison with industrial sector. In today's world consumer boom, essential oils are exposing hitherto unexplored areas of industrial applications. Some of the emerging areas are the following.

Biocides

Herbs and their constituents manifest varying degrees of pest controlling activities like antimicrobial, antifungal, insecticidal, ovicidal, attractant, repellent, antifeedant and antigonadal. Thus, they are a potent source of environmentally and ecologically safe pesticides and could be exploited for commercialisation. Biocide preparations from herbs and extraction and production of plant growth regulators and allelochemicals are areas where our indigenous knowledge and bio-diversity resources can be profitably exploited. Failures and non-sustainability of chemical route to agriculture and healthcare provide an opportunity to re-evaluate our traditional knowledge systems on safe use of plant extracts for pest and disease management. Growing awareness of environment hazards from synthetic pesticides and associated problems of pest resistance and detrimental effects on non-target organisms dictate need for safe, effective and economical pesticides. Several components of essential oils are identified to possess short term (linalool, menthol, pulegone) and long term (benzaldehyde, carvacrol, carvone, cinnamaldehyde) fungistatic effects. Global agro-environmental concerns necessitated search for safer and viable alternatives to hazardous synthetic pesticides. Secondary metabolites such as essential oils are inexpensive, biodegradable, more systemic and environment friendly botanical pesticides. Use of botanicals may prove to be one of the best ways to manage food losses by pests in future.

Allelochemicals

A number of plant species produce volatile terpenes that can influence nearby growing plants. A variety of allelochemicals is identified including volatile terpenes. Allelochemicals are resources for developing herbicides and plant growth stimulants. *Salvia sclaria* produces volatile inhibitors such as camphor, 1,8-cineole, pinene and dipentene, which inhibit DNA synthesis, seed germination and growth of many weeds. Germination inhibiting activity is reported for essential oils of peppermint, ajowan, basil and spearmint. Monoterpenoids such as carvone, linalool, methyl chavicol and anethol are thus most effective.

Storage and preservation

Essential oils and herbs are useful against fungal and insect infestation of various stored food commodities. Volatile oils of *Pongamia glabra, Acorus calamus, Cedrus deodora, Citrus medica, Ocimum canum, O. adscendens,* and *O. gratissimum* are potent agents against storage pests and fungal pathogens. Essential oil constituents such as citral, citronellal, citronellol, eugenol, farnesol and nerol are potential inhibitors of fungal infestation. Decay and sprouting of onion bulbs are suppressed by oils of eucalyptus, lemon grass and camphor and also by citronellol and thymol. Essential oil has sprout-inhibiting property on potatoes too.

Aromatherapy

Aromatherapy is an emerging art of healing using essential oils, which have therapeutic effect on body, mind and soul. Alternative and relatively less costly health care system of aromatherapy offers therapeutic effects leading to calmness, emotional balance, stress relief and rejuvenation. Essential oils have a variety of therapeutic activities including antiseptic, sedative, antiinflammatory, antispasmodic, antifungal, antiviral, stimulant, relaxant, diuretic, invigorating, euphoric and digestive. India has a long tradition of using aromatic and medicinal plants in its holistic medicinal systems- *Ayurveda, Unani and Sidha.* Aromatherapy was brought into practice when antiseptic and skin permeability properties of essential oils were discovered. Methods used in aromatherapy are inhalation, local application and baths, which permit essential oils to enter the body. Once within the body, essential oils re-establish a close harmony with affected body parts and revitalise them. Different blends of essential oils are used for various ailments like muscular aches and pains, respiratory problems, insomnia, headache, swollen joints, urinary disorders, skin infections, palpitation, depression, indigestion, gynaecological problems, etc. Unani system of medicine employs more than 40 aromatic and medicinal plants containing essential oils as their principal constituents, such as *Lavandula stoechas* (ustukhudoos), *Pimpinella anis* (anisoon), *Artemisia absenthium* (afsanteen), *Aqualaria agalocha* (agar), *Balsomodendron opobalsamum* (balsam) and *Anethum sowa* (sowa). This can easily be integrated with eco-tourism, which helps to find domestic market for indigenous essential oils thus gearing up cultivation and production sectors.

Value addition

Post harvest processing of aromatic and medicinal materials and value addition to products of cultivated aromatic and medicinal plants are receiving attention of biochemists, phytochemists and chemical engineers who are designing effective and efficient equipments for cold and hot extraction of oils, fractions and crude extracts for domestic use and export. Possibilities on use of microorganisms to produce some of value added materials bio-technologically are opening up. Indian industry also needs to perfect use of appropriate technologies for fractionation of major and minor components of high value from bulk-produced essential oils of low sale value and transformation of low value components into high value aroma products. Encapsulated spice extractives extended on a salt dextrose carrier, encapsulated spice oils and oleoresins and homogenous free flowing oleoresins are to be commercialised. With emergence of 'nature food', 'ethnic food' and 'yogic food' and emphasis on 'back to nature', uses of spice based oils, oleoresins, ointments and flavourings would be in a bullish market in the new millennium.

New molecules

Currently about 150 aromatic and medicinal plants and 70 spices are cultivated in different parts of the world. Out of these, probably 50 aromatic and medicinal plants and 35 spices are commonly used. But there are about 2500 Indian plant species belonging to about 60 families, which have aromatic and medicinal principles. Many of these species could offer new essential oils and aroma chemicals to meet ever changing and ever growing demand for novel natural essential oils and aroma chemicals. Potential of such lesser-known aromatic and medicinal plants needs further studies and exploitation. Medicinal and aromatic plants are known as the biological mine of novel products and applications. They produce a variety of biologically active metabolites, which may confer them competitive advantage to fight against biotic stresses. Many a time, minor components in plant material generate great demand. Hence, identification of minor chemical constituents and developing chemical profile is a felt need. Species like *Saussurea costus, Nardostachys grandiflora, Pinus roxburghii, Piper nigrum, Skimmia laureola, Zanthoxylum armatum, Cedrus deodara, Cinnamomum tamala, Curcuma angustifolia, C. zedoaria, Alpinia galanga, A. calcarata* and *Kaempferia galanga* contain essential oils in various parts.

Eco-tourism

Australia is the first country to have an eco-tourism strategy and Malaysia has followed suit. Diverse ecological zones of India offer scope for developing eco-tourism in the country. Development of eco-tourism can generate substantial foreign exchange without the usual disastrous ecological degradation associated with general tourism. It will help to promote and popularize various indigenous systems of medicine, take people closer to nature and above all promote overall well being of people.

Eco-technology

Ecological foundations essential for sustained advances in biological productivity and the atmosphere are experiencing gradual degradation. There is much to learn from the past in terms of ecological and social sustainability of technologies. Developments have opened new opportunities for developing technologies, which can lead to high productivity without any adverse effect on natural resource base. Blending traditional and frontier technologies lead to birth of eco-technologies with combined strength of economics, ecology, equity, energy and employment. There is need to conserve traditional wisdom and practices which are often tending to become extinct. Eco-technologies enable adoption of ISO 9000 and ISO 14000 standards of environmental management.

Conclusion

Rich bio-diversity of plants and diverse knowledge systems in harnessing plant bio-diversity provide an opportunity to meet future challenges in agriculture, health care systems, fragrance, flavors and allied areas. Failures and non-sustainability of so-called modern approaches to agriculture and health care systems could be re-assessed through our knowledge heritage and natural resources. If information technology holds today's fate, biotechnology determines the future. A strategic team effort in aromatic plants research and development will pave way for reaping the *green gold* for which India is the richest repository.

References

Atal, C.K. and Kapur, B.M. 1997. *Cultivation and Utilization of Medicinal and Aromatic Plants.* Regional Research Laboratory, Jammu-Tawi, India, 815p

Balz, R. 1999. *The Healing Power of Essential Oils.* Motilal Banarsidass Publishers Pvt Ltd, Delhi, 203p

Farooqi, A.A. and Sreeramu B.S. 2001. *Cultivation of Medicinal and Aromatic Crops.* Universities Press (India) Ltd., Hyderabad, 518p

Guenther, E. 1950. *The Essential Oils.* Vol. IV. Van Nostrand Co. Inc., New York, 752p

Handa, S.S. and Kaul, M.K. 2001. *Supplement to Cultivation and Utilization of Aromatic Plants.* Regional Research Laboratory, Jammu-Tawi, 352p

Husain, A., Virmani, O.P., Sharma, A., Kumar, A. and Misra, L.N. 1988. *Major Essential Oil Bearing Plants of India.* Central Institute of Medicinal and Aromatic Plants, Lucknow, India. 237p

Joy, P.P., Thomas, J., Mathew, S., Jose, G. and Joseph, J. 2001. Aromatic plants. In: *Tropical Horticulture* Vol. 2. Bose T.K., Kabir J., Das P. and Joy P.P.(eds.), Naya Prokash, Calcutta, India, pp 633-733

Kurian, A. and Peter, K.V. 2005. Aromatic plants in pharmaceuticals. In: *Proc. Vaidyaratnam Global Summit on Ayurveda & Expo 05, 9-11 Dec.2005,* Thrissur. Seminar Abstracts and Papers, Ashtavaidyan Thaikkattu Mooss, Vaidyaratnam Oushadhasala, Thrissur, India, pp 74-85

Pareek, S.K. and Gupta, R. 1985. On the status of agronomic research in *Cymbopogon* grasses in India with projections on future work. Indian Perfumer, 29: 215-224

Thomas, J. 1995. Lemongrass. In: *Advances in Horticulture Vol. II- Medicinal and Aromatic Plants*. Chadha, K.L. and Gupta, R. (eds.), Malhotra Publishing House, New Delhi, 726p

Weiss, E.A. 1997. *Essential Oil Crops*. CAB International, New York, 600p

2

Classification of Aromatic Plants

Though more than 1500 species contain volatile aromatic principles, only around 50 species serve as commercial source of essential oil and aroma-chemicals. Among these, aromatic crops under large-scale cultivation hardly exceeds two dozens. Important aromatic plants of world are classified according to economic part used, growth habit, habitat, crop duration and propagation method besides usual botanical classification.

1. *Based on importance*

Plants grown exclusively for extraction of aromatic principles for use in perfumery and cosmetics are classified as major whereas those, in which volatile oil and aroma principles are byproducts or secondary products, are classified as minor aromatic crops.

i *Major aromatic crops:* Chamomile, Vetiver, Lemongrass, Patchouli, Tea tree, Eucalyptus

ii *Minor aromatic crops:* Cinnamon, Marigold, Dill, Ambrette, Celery

2. *Based on economic part*

i *Herbage:* Patchouli, Citronella, Sweet basil, Geranium, Rosemary

ii *Root:* Vetiver, *Sassafras albidum*, Sandalwood, Camphor

iii *Wood: Ocotea cymbarum, Ocotea pretiosa,* Sandalwood, camphor, Linaloe.

iv *Bark: Cinnamomum verum, C. burmannii*

v *Leaf:* Eucalyptus, Tea tree, *Skimmia laureola, Doryphora sassafras, Laurus nobilis,* Lemon grass, Mint, Camphor

vi *Flower:* Jasmine, Rose, Marigold, Chamomile, Champak, Tuberose, Ylang ylang

vii *Flowering tops:* Davana, Palmarosa, Thyme

viii *Fruit:* Dill, *Litsea cubeba,* Linaloe

ix *Seed:* Ambrette, Ajowan, Celery, Clarysage

3. Based on growth habit

i *Grasses:* Palmarosa, Rosha grass, Lemongrass, Vetiver, Citronella

ii *Herbs:* Sweet basil, Tuberose, Thyme, Rosemary, Chamomile, Ajowan, Davana, Marigold, Mint

iii *Shrubs: Skimmia laureola,* Patchouli, Rose, Geranium, Jasmine

iv *Trees:* Eucalyptus, Tea tree, Camphor, Champak, Cinnamon, Linaloe, Ylang ylang

4. Based on habitat

i *Tropical:* Lemon grass, Ocimum, Cinnamon, Linaloe, Sandalwood, Eucalyptus, Citronella, Palmarosa, Patchouli, Vetiver, Ylang ylang,

ii *Sub-tropical:* Vetiver, Mint, Eucalyptus, Ajowan, Thyme, Rosemary, Citronella, Davana, Fennel, Japanese mint

iii *Temperate:* Chamomile, Ajowan, Fennel, Pepper mint, Spear mint, Bergamot mint

5. Based on crop duration

i *Annuals:* Chamomile, *Ocimum basilicum,* Ajowan, Davana

ii *Biennials:* Vetiver, Celery

iii *Perennials:* Lemon grass, Geranium, Lemon grass, Mint, Palmarosa, Rose, Cinnamon, Ylang ylang, Tea tree

6. Based on method of propagation

i *Vegetatively propagated: Citronella, Geranium, Jasmine, Patchouli, Rose, Tuberose*

ii *Sexually (seed) propagated:* Clarysage, Cumin, Davana, Camphor, Eucalyptus, Sandalwood, Palmarosa, Ylang Ylang

iii *Both vegetatively and sexually propagated:* Lemon Grass, Linaloe, Marigold, Palmarosa, Rosemary, Thyme, Vetiver

7. Botanical classification:

DIVISION : EMBRYOPHYTA

Subdivision: Gymnospermae

Class: Coniferae

Family	Genus	Species
Podocarpaceae	*Dacrydium*	*Dacrydium franklini*
Pinaceae	*Picea*	*Picea abies*
		P. alba
		P. canadensis
		P. excelsa
		P. glauca
		P. jezoensis
		P. mariana
		P. nigra
		P. obovata
		P. vulgaris
	Tsunga	*Tsuga canadensis*
		T. douglasii
		T. heterophylla
	Pseudotsuga	*Pseudotsuga douglasii*
		P. glauca
		P. mucronata
		P. taxifolia
	Abies	*Abies alba*
		A. balsamea
		A. balsamifera
		A. douglasii
		A. excels
		A. mayriana
		A. mucronata
		A. pectinata
		A. picea
		A. sachalinensis
		A. sibirica
	Cedrus	*Cedrus atlantica*
		C. deodara
		C. libani
		C. libanotica
	Pinus	*Pinus albicaulis*
		P. aristata
		P. attenuata
		P. ayacahuite
		P. balfouriana
		P. balsamea
		P. banksiana
		P. caribaea
		P. cembra

		P. clausa
		P. contorta
		P. coulteri
		P. echinata
		P. edulis
		P. flexilis
		P. glabra
		P. jeffreyi
		P. lambertiana
		P. longifolia
		P. monophylla
		P. montana
		P. monticola
		P. mugo
		P. roxburghi
Taxodiaceae	*Sciadopitys*	*Sciadopitys verticillata*
	Cryptomeria	*Cryptomeria japonica*
Cupressaceae	*Callitropsis*	*Callitropsis araucarioides*
	Thujopsis	*Thujopsis dolabrata*
	Thuja	*Thuja plicata*
	Cupressus	*Cupressus fastigiata*
		C. glauca
		C. japonica
		C. lambertiana
		C. lawsoniana
		C. lusitanica
		C. macrocarpa
		C. pendula
		C. sempervirens
		C. sinensis
		C. torulosa
	Chamaecyparis	*Chamaecyparis lawsoniana*
		C. obtusa
		C. taiwanensis
		C. thyoides
	Juniperus	*Juniperus communis*
		J. mexicana
		J. oxycedrus
		J. phoenicea
		J. procera
		J. sabina
		J. thurifera
		J. virginiana

Subdivision: Angiospermae
Class: Monocotyledonae

Family	Genus	Species
Graminae (Poaceae)	*Elyonurus*	*Elyonurus latiflorus*
		E. tripsacoides
	Vetiveria	*Vetiveria zizanioides*
	Cymbopogon	*Cymbopogon afronardus*
		C. caesius
		C. citratus
		C. clandestinus
		C. coloratus
		C. confertiflorus
		C. densiflorus
		C. exaltatus
		C. flexuosus
		C. georingii
		C. giganteus
		C. jwarancusa
		C. rectus
		C. martinii
		C. nardus
		C. nervatus
		C. polyneuros
		C. procerus
		C. proximus
		C. schoenanthus
		C. senaarensis
		C. stipulatus
		C. virgulus
		C. winterianus
	Andropogon	*Andropogon aciculatus*
		A. connatus
		A. fragrans
		A. intermedius
		A. kuntzeanus
		A. muricatus
		A. nardoides
		A. odoratus
		A. versicolor
Cyperaceae	*Cyperus*	*Cyperus rotundus*
Araceae	*Acorus*	*Acorus calamus*
Pandanaceae	*Pandanus*	*Pandanus fascicularis* (Syn: *P. odoratissimus*)
		P. tectorius

Arecaceae	*Cocos*	*Cocos nucifera*
Liliaceae	*Allium*	*Allium cepa*
		A. sativum
	Lilium	*Lilium candidum*
	Hyacinthus	*Hyacinthus non-scriptus*
		H. orientalis
	Convallaria	*Convallaria majalis*
Amaryllidaceae	*Narcissus*	*Narcissus jonquilla*
		N. poeticus
		N. tagetta
	Polyanthes	*Polyanthes tuberosa*
Iridaceae	*Crocus*	*Crocus sativus*
	Iris	*Iris florentina*
		I. germanica
		I. pallida
Zingiberaceae	*Hedychium*	*Hedychium flavum*
	Kaempferia	*Kaempferia galanga*
		K. rotunda
	Curcuma	*Curcuma amada*
		C. aromatica
		C. caesia
		C. domestica
		C. longa
		C. xanthorrhiza
		C. zedoaria
		C. zerumbet
	Alpinia	*Alpinia alleghas*
		A. galanga
		A. officinarum
		A. calcarata
		A. khulanjan
		A. malaccensis
		A. nutans
	Zingiber	*Zingiber mioga*
		Z. nigrum
		Z. officinale
		Z. elatum
		Z. zerumbert
		Z. cassunar
	Amomum	*Amomum angustifolium*
		A. aromaticum
		A. cardamom
		A. globosum
		A. hirsutum

		A. korarima
		A. melegueta
		A. subulatum
	Elettaria	*Elettaria cardamomum*

Class: Dicotyledonae

Piperaceae	*Piper*	*Piper acutifolium*
		P. angustifolium
		P. asperifolium
		P. camphoriferum
		P. clusii
		P. crassipes
		P. cubeba
		P. guineense
		P. lineatum
		P. longum
		P. lowong
		P. mollicomum
		P. molissimum
		P. nigrum
		P. officinarum
		P. ribesioides
		P. betle
		P. pinnatum
		P. boehmeriaefolium
Betulaceae	*Betula*	*Betula alba*
		B. brea
		B. dulce
		B. lenta
		B. papyrifera
		B. pendula
		B. pubescens
Moraceae	*Humulus*	*Humulus americanus*
		H. lupulus
Santalaceae	*Osyris*	*Osyris tenuifolia*
	Santalum	*Santalum album*
		S. lanceolatum
		S. preissianum
		S. spicatum
		S. zygnorum
		S. acuminatum
		S. murrayanum
	Fusanus	*Fusanus spicatus*

Aristolochiaceae	*Asarum*	*Asarum canadense*
		A. europaeum
Chenopodiaceae	*Chenopodium*	*Chenopodium ambrosioides*
Myristicaceae	*Myristica*	*Myristica argentea*
		M. fragrans
		M. malabarica
		M. succedanea
		M. canarica
		M. laurifotia
		M. muelleri
		M. elliptica
		M. castaneaefolia
Lauraceae	*Cinnamomum*	*Cinnamomum aromaticum*
		C. burmannii
		C. camphora
		C. cassia
		C. cecidodaphne
		C. culilawan
		C. glanduliferum
		C. impressinervium
		C. iners
		C. kanahirai
		C. loureirii
		C. micranthum
		C. obtusifolium
		C. tamala
		C. xanthoneuron
		C. zeylanicum (*C. verum*)
	Ocotea	*Ocotea caudata*
		O. cymbarum
		O. parviflora
		O. pretiosa
		O. bullata
		O. radiance
		O. wrightii
	Sassafras	*Sassafras albidum*
		S. randainese
	Cryptocaria	*Cryptocaria massoia*
	Laurus	*Laurus nobilis*
		L. azorica
		L. cubeba
		L. diversiflora
	Umbellularia	*Umbellularia californica*

	Aniba	*Aniba parviflora*
		A. rosaeodora
	Cryptocarya	*Cryptocarya massoia*
	Ravensara	*Ravensara aromatica*
	Dicypellium	*Dicypellium caryophyllatum*
Monimiaceae	*Doryphora*	*Doryphora sassafras*
Ranunculaceae	*Nigella*	*Nigella damascena*
Magnoliaceae	*Magnolia*	*Magnolia grandiflora*
		M. soulangiana
		M. coco
	Michelia	*Michelia champaca*
		M. longifolia
		M. excelsa
		M. figo
		M. kisopa
		M. nilagirica
		M. rheedi
	Illicium	*Illicium anisatum*
		I. japonicum
		I. religiosum
		I. verum
Anonaceae	*Cananga*	*Cananga odorata*
		C. latifolia
Cruciferae	*Cochlearia*	*Cochlearia armoracia*
	Brassica	*Brassica alba*
		B. juncea
		B. napus
		B. nigra
	Raphanus	*Raphanus sativus*
Violaceae	*Viola*	*Viola odorata*
Flacourtiaceae	*Oncoba*	*Oncoba spinosa*
Caryophyllaceae	*Dianthus*	*Dianthus caryophyllus*
Resedaceae	*Reseda*	*Reseda odorata*
Clusiaceae	*Calophylum*	*Calophylum inophyllum*
(Guttiferae)	*Mesua*	*Mesua ferra*
	Ochrocarpus	*Ochrocarpus longifolius*
Dipterocarpaceae	*Dryobalanops*	*Dryobalanops aromatica*
		D. camphora
	Dipterocarpus	*Dipterocarpus tuberculatus*
		D. turbinatus
Saxifragaceae	*Philadelphus*	*Philadelphus coronarius*
Hamamelidaceae	*Hamamelis*	*Hamamelis virginiana*
	Liquidambar	*Liquidambar orientalis*
		L. styraciflua

Rosaceae	*Spiraea*	*Spiraea ulmaria*
	Rosa	*Rosa alba*
		R. canina
		R. centiflolia
		R. damascena
		R. gallica
		R. indica
		R. glandulifera
		R. moschata
		R. pubescens
		R. multiflora
		R. chinensis
		R. rugosapp
		R. pannonica
		R. gigantica
	Prunus	*Prunus amygdalus*
		P. laurocerasus
Leguminosae	*Acacia*	*Acacia cavenia*
		A. dealbata
		A. decurrens
		A. farnesiana
		A. floribunda
	Copaifera	*Copaifera coriacea*
		C. glycycarpa
		C. guianensis
		C. martii
		C. multijuga
		C. officinalis
		C. reticulata
	Myroxylon	*Myroxylon balsamum*
		M. pereirae
	Lupinus	*Lupinus luteus*
	Genista	*Genista sibirica*
		G. tinctoria
	Spartium	*Spartium junceum*
	Wistaria	*Wistaria sinensis*
	Hardwickia	*Hardwickia mannii*
	Myrocarpus	*Myrocarpus fastigiatus*
		M. frondosus
	Parkinsonia	*Parkinsonia aculeata*
	Glycyrrhiza	*Glycyrrhiza glabra*
Geraniaceae	*Geranium*	*Geranium lugubre*
		G. macrorrhizum
		G. maculatum

		G. robertianum
	Pelargonium	*Pelargonium capitatum*
		P. fragrans
		P. graveolens
		P. odoratissimum
		P. radula
		P. roseum
		P. terebinthinaceum
Rutaceae	*Murraya*	*Murraya exotica*
		M. paniculata
	Skimmia	*Skimmia laureola*
	Xanthoxylum	*Xanthoxylum piperitum*
	Ruta	*Ruta angustifolia*
		R. bracteosa
		R. graveolens
		R. montana
	Pilocarpus	*Pilocarpus jaborandi*
		P. microphyllus
		P. racemosus
		P. spicatus
	Cusparia	*Cusparia trifoliata*
	Boronia	*Boronia megastigma*
	Barosma	*Barosma betulina*
		B. crenulata
		B. serratifolia
	Amyris	*Amyris balsamifera*
	Clausena	*Clausena medico*
		C. anisata
		C. anisum-olens
	Citrus	*Citrus acida*
		C. aurantifolia
		C. aurantium
		C. decumana
		C. deliciosa
		C. excavata
		C. latifolia
		C. limetta
		C. limon
		C. maxima
		C. medica
		C. nobilis
		C. paradisi
		C. reticulata
		C. sinesis

		C. unshiu
Burseraceae	*Boswellia*	*Boswellia carterii*
		B. serrata
	Bursera	*Bursera aloexylon*
		B. delpechiana
		B. fragroides
		B. glabrifolia
	Commiphora	*Commiphora abyssinica*
		C. erythraea
		C. myrrha
		C. schimperi
	Canarium	*Canarium luzonicum*
Meliaceae	*Melia*	*Melia azadiracta*
Euphorbiaceae	*Croton*	*Croton eluteria*
Anacardiaceae	*Pistacia*	*Pistacia lentiscus*
	Schinus	*Schinus molle*
Moringaceae	*Moringa*	*Moringa oleifera*
Tiliaceae	*Tilea*	*Tilea cordata*
		T. platyphyllos
		T. tomentosa
Malvaceae	*Hibiscus*	*Hibiscus abelmoschus*
Zygophyllaceae	*Bulnesia*	*Bulnesia sarmienti*
Sterculiaceae	*Dombeya*	*Dombeya viscose*
	Cola	*Cola acuminata*
Cistaceae	*Cistus*	*Cistus ladaniferus*
Myrtaceae	*Myrtus*	*Myrtus aeris*
		M. caryophyllata
		M. communis
		M. pimenta
	Pimenta	*Pimenta acris*
		P. citrifolia
		P. officinalis
		P. racemosa
	Eugenia	*Eugenia acris*
		E. caryophyllata
		E. pimenta
		E. brasiliensis
		E. uniflora
		E. jambolana
	Syzygium	*Syzygium cumini*
	Leptospermum	*Leptospermum citratum*
		L. flavescens
	Melaleuca	*Melaleuca alternifolia*
		M. bracteata

	M. cajeputi
	M. leucodendron
	M. linariifolia
	M. maideni
	M. minor
	M. quinquenervia
	M. smithii
	M. trichyostachya
	M. viridiflora
Eucalyptus	*Eucalyptus acmenoides*
	E. amygdalina
	E. astringents
	E. australiana
	E. behriana
	E. bicostata
	E. camaldulensis
	E. citriodora
	E. cneorifolia
	E. crebra
	E. deylupta
	E. dives
	E. dumosa
	E. eginoides
	E. elaeophora
	E. elata
	E. fruticetorum
	E. globulus
	E. grandis
	E. hemipholia
	E. leucoxylon
	E. lindleyana
	E. macarthuri
	E. maculosa
	E. maidenii.
	E. melanopholia
	E. melliodora
	E. numerosa
	E. oleosa
	E. pachyphylla
	E. paniculata
	E. phellandra
	E. pilularis
	E. piperita
	E. polybractea

		E. propinqua
		E. punctata
		E. pyconantha
		E. radiata
		E. robusta
		E. sideroxylon
		E. smithii
		E. staigerana
		E. tereticornis
		E. torelliana
		E. viridis
		E. youmann
Lythraceae	*Lawsonia*	*Lawsonia alba*
Combretaceae	*Quisquaqlis*	*Quisquaqlis indica*
Umbelliferae	*Coriandrum*	*Coriandrum sativum*
(Apiaceae)	*Cuminum*	*Cuminum cyminum*
	Apium	*Apium graveolens*
		A. petroselinum
	Petroselinum	*Petroselinum hortense*
		P. sativum
	Carum	*Carum ajowan*
		C. bulbocastanum
		C. carvi
		C. copticum
		C. petroselinum
		C. verticillatum
	Pimpinella	*Pimpinella anisum*
		P. diversifolia
		P. saxifragra
	Foeniculum	*Foeniculum vulgare*
	Anethum	*Anethum graveolens*
		A. sowa
	Oenanthe	*Oenanthe phellandrium*
	Levisticum	*Levisticum officinale*
	Angelica	*Angelica archangelica*
		A. atropurpurea
		A. glabra
		A levisticum
	Ferula	*Ferula alliacea*
		F. asafetida
		F. badra-kema
		F. ceratophylla
		F. foetida
		F. galbaniflua

		F. rubricaulis
		F. suaveolens
		F. sumbul
	Peucedanum	*Peucedanum ostruthium*
	Daucus	*Daucus carota*
	Crithmum	*Crithmum maritimum*
Primulaceae	*Cyclamen*	*Cyclamen europaeum*
Oleaceae	*Syringa*	*Syringa vulgaris*
	Jasminum	*Jasminum officinale*
		J. grandiflorum
		J. auriculatum
		J. sambac
		J. undulatum
		J. paniculatum
		J. nitidum
		J. rigidum
		J. arborescens
		J. flexile
		J. pubescens
	Calophulum	*Calophulum communis*
		C. indulatum
		C. rotterianum
		C. multiflorum
		C. malabaricum
		C. rigidum
	Osmathus	*Osmathus fragrans*
		O. americanus
	Nyctanthes	*Nyctanthes arbortristris*
Apocynaceae	*Nerium*	*Nerium odorum*
	Tarbernaemontana	*Tarbernaemontana coronaria*
	Thevetia	*Thevetia nerifolia*
Verbenaceae	*Lippia/ Aloysia*	*Lippia citriodora*
	Clerodendron	*Clerodendron inerme*
Solanaceae	*Capsicum*	*Capsicum minimum*
		C. fruitescens
		C. annuum
Bignoniaceae	*Millingtonia*	*Millingtonia hortensis*
Labiatae (Lamiaceae)	*Rosmarinus*	*Rosmarinus flexuosus*
		R. lavandulaceus
		R. laxiflorus
		R. officinalis
		R. tournefortii
	Lavandula	*Lavandula barmanni*
		L. dentate

	L. hybrida
	L. intermedia
	L. latifolia
	L. officinalis
	L. pedunculata
	L. spica
	L. stoechas
	L. vera
	L. viridis
Nepeta	*Nepeta cataria*
	N. liniaris
	N. spicata
Salvia	*Salvia carnosa*
	S. espanola
	S. hiemalis
	S. hispanorum
	S. lavandulaefolia
	S. leucophylla
	S. muscatel
	S. officinalis
	S. sclarea
	S. triloba
	S. verbenaea
Monarda	*Monarda citriodora*
	M. fistulosa
	M. menthaefolia
	M. pectinata
	M. punctata
Melissa	*Melissa officinalis*
Hedeoma	*Hedeoma pulegioides*
Satureia	*Satureia hortensis*
	S. montana
Hyssopus	*Hyssopus officinalis*
Origanum	*Origanum compactum*
	O. elongatum
	O. fort-queri
	O. grossi
	O. majorana
	O. virens
	O. vulgare
Marjorana	*Marjorana silvestre*
	M. hortensis
Thymus	*Thymus capitatus*
	T. cephalotus

	T. hiemalis
	T. hirtus
	T. loscossi
	T. mastichina
	T. serpyllum
	T. virginicus
	T. vulgaris
	T. zygis
Mentha	*Mentha aquatica*
	M. arvensis
	M. cablin
	M. canadensis
	M. citrata
	M. japonica
	M. longifolia
	M. piperita
	M. pulegium
	M. rotundifolia
	M. spicata
	M. sylvestris
	M. verticillata
	M. viridis
Perilla	*Perilla citriodora*
	P. frutescens
	P. nankinensis
	P. ocymoides
Pogostemon	*Pogostemon cablin* (*syn:P. patchouli*)
	P. heyneanus
	P. hortensis
	P. comosum
	P. plectranthoides
Ocimum	*Ocimum americanum*
	O. basilicum
	O. canum
	O. carnosum
	O. gratissimum
	O. kilimandscharicum
	O. album
	O. anisatum
	O. menthaefolium
	O. micranthum
	O. minimum
	O. nakurense

		O. pilosum
		O. sanctum
		O. suave
		O. viride
	Mosla/Orthodon	*Mosla angustifolia*
		M. chinesis
		M. formosana
		M. hadai
		M. japonica
		M. lanceolata
		M. lysimachiiflora
		M. punctata
		M. thymolifera
	Pycnanthemum	*Pycnanthemum incanum*
		P. lanceolatum
		P. muticum
		P. pilosum
	Coridothymus	*Coridothymus capitatus*
Myoporaceae	*Eremophila*	*Eremophila mitchelli*
Rubiaceae	*Gardenia*	*Gardenia citriodora*
		G. florida
		G. grandiflora
		G. longistyla
		G. resinifera
		G. floribunda
		G. latifolia
	Leptactina	*Leptactina senegambica*
	Hamiltonoa	*Hamiltonoa suaveolens*
	Cinchona	*Cinchona succirubra*
		C. ledgeriana
		C. calisaya
	Hydychium	*Hydychium spicatum*
Ericaceae	*Gaultheria*	*Gaultheria procumbens*
		G. fragrantissima
Sapotaceae	*Mimusops*	*Mimusops elengi*
Caprifoliaceae	*Lonicera*	*Lonicera caprifolium*
		L. gigantea
		L. japonica
Valerianaceae	*Valeriana*	*Valeriana celfica*
		V. officinalis
		V. wallichii
		V. brunoniana
		V. hardwickii

Compositae (Asteraceae)	*Solidago*	*Solidago odora*
	Erigeron	*Erigeron canadensis*
	Blumea	*Blumea balsamifera*
		B. lacera
		B. ampletectens
		B. densiflora
		B. aurita
		B. glabra
	Helichrysum	*Helichrysum angustifolium*
		H. arenarium
		H. italicum
		H. stoechas
	Inula	*Inula helenium*
	Tagetes	*Tagetes glandulifera*
		T. minuta
		T. erecta
		T. patula
		T. lucida
	Santolina	*Santolina chamaecyparissus*
	Anthemis	*Anthemis nobilis*
	Achillea	*Achillea millefolium*
		A. moschata
	Matricaria	*Matricaria chamomilla*
		M. inodora
	Artemisia	*Artemisia absinthium*
		A. cina
		A. dracuculus
		A. maritima
		A. pallens
		A. pontica
		A. tridentata
		A. vulgaris
		A. vestita
		A. scoparia
		A. parviflora
		A. pollens
	Arnica	*Arnica montana*
	Saussurea	*Saussurea lappa*
	Tanacetum	*Tanacetum vulgare*

References

Atal, C.K and Kapur, B.M. 1982. *Cultivation and Utilization of Aromatic Plants*. Regional Research Laboratory, Jammu-Tawi, 815p

Joy, P.P., Thomas, J., Mathew, S., Jose, G. and Joseph, J. 2001. Aromatic plants. In: *Tropical Horticulture* Vol. 2. Bose T.K., Kabir J., Das P. and Joy P.P.(eds.), Naya Prokash, Calcutta, India, pp 633-733

Kapoor, L.D. and Krishnan, R. 1977. *Advances in Essential Oil Industry*. Today and Tomorrow's Printers, New Delhi, 245p

Mahindru, S.N. 1992. *Indian Plant Perfumes.* Metropolitan, New Delhi, 263p

Weiss, E.A. 1997. *Essential Oil Crops.* CAB International, New York, 600p

3

Extraction of Aroma Principles

Essential, volatile or ethereal oils are mixtures composed of volatile liquid and solid compounds which vary widely in regard to their composition and boiling points. Plants owe their fragrance to presence of traces of essential oils in different parts. Numerous fragrance materials are present in roots, stems, barks, leaves, flowers, fruits and heartwoods.

Several processes like distillation, enfleurage, maceration, expression, solvent extraction and fluid extraction are available for extraction of aroma principles. Application of these processes, either singly or in combination depends upon nature of material and essential oil or absolute intended to be recovered.

I. Distillation

Distillation may be defined as separation of components of a mixture of two or more liquids by virtue of difference in their vapour pressure.

The bulk of essential oils are produced by distillation. There are three systems of distillation- hydro, hydro-steam and steam distillation.

a. Hydrodistillation

Hydrodistillation though the oldest, is still being widely practised for oil extraction. The plant material is in direct contact with boiling water in a crude metallic distillation unit. The material floats on water or be completely immersed, depending upon its specific gravity and quantity of material handled per charge. Water is boiled by direct fire, steam jacket, closed steam coil or in a few cases open or perforated steam coil. Some plant materials like powdered almonds, rose petals and orange blossoms must be distilled while fully immersed and moving freely in boiling water, because on distillation with injected live steam (direct steam distillation) these materials agglutinate and form large compact lumps through which steam cannot penetrate.

b. Hydro-steam distillation

Hydro-steam distillation is employed where perfumery material is vulnerable to direct steam. Consequently, plant material is supported on a perforated grid or screen inserted at some distance above bottom of still. Lower part of still contains water up to a level just below grid. Water may be heated by any of the methods. Wet steam of low pressure rises through plant material. Typical features of this method are that steam is always fully saturated, wet and never superheated; besides plant material is in contact with steam only and not with boiling water.

c. Steam distillation

In steam distillation, live steam saturated or superheated, under pressure (up to 7 kg/cm^2) is injected through steam tubes below the charge and pressure within the distillation vessel is controlled according to nature of the material being distilled. It resembles hydro-steam distillation except that no water is kept in bottom of still. This method is efficient and gives higher yields. However, it is not generally employed to delicate flowers.

Equipment for distillation

Equipment required for carrying on distillation of plant materials depends upon the size of operation and type of distillation to be used. There are, however three main parts which in varying size, form base for all three types of hydro distillation. The three universally employed parts are retort or still proper, condenser and oil separator. A fourth part namely, a separate steam boiler is necessary for steam distillation wherein direct live steam, often slightly superheated is required.

Retort: Retort or still proper commonly called as 'tank' serves primarily as a container for plant material and as a vessel in which water and or steam contacts plant material and vaporizes its essential oil.

Condenser: Condenser serves to convert all of steam and accompanying oil vapours into liquid.

Oil separator: Its function is to achieve a quick and complete separation of oil from condensed water. Since, total volume of water condensed will always be much greater than quantity of oil, it is necessary to remove the water continuously. Condensate flows from condenser into oil separator, where distillation water and volatile oil separate automatically.

II. **Maceration** (Extraction with hot fat)

In maceration, oil cells of fragrant flowers are ruptured by immersion in a hot fat or oil at 60-70°C which in turn absorbs essential oils. Fat is separated from spent flowers and reused for absorbing fragrance from next batch of fresh flowers. Fat retained by flowers is recovered by hydraulic pressing. Resultant perfumed pomade is frequently marketed as such but is often extracted with strong alcohol to yield extracts. These alcoholic absolutes are absolutes of pomade in market parlance.

III. **Enfleurage** (Extraction with cold fat)

Enfleurage is the process of extraction of fragrance by absorbing it from flowers in contact with cold fats. This process is adopted for fragrant flowers of jasmine and tuberose, which continue to manifest their characteristic fragrance even in plucked condition. Solvents lack this virtue of arresting manifested fragrance.

Fats should be saturated and odourless to prevent entrance of fat odours. Refined lard or beef suet are preferred. Fat is thinly layered on both sides of a glass plate supported on a rectangular wooden frame or chassis. Fresh fragrant flowers are lightly layered on fat coated chassis. Several chassis are placed one above the other sandwiching the flowers between two layers of fat. Spent flowers are removed (defleurage) and fresh charge is made. Reversing of glass slab is called *patage*. Patage is done several times to obtain maximum perfume absorption. Furrows are created with combs to increase absorption surface. The process of defleurage, fresh charging and patage is continued to obtain fat of desired perfume strength.

Enfleurage gives a much greater yield of flower oil than other methods. Despite this advantage, enfleurage has lately been replaced by extraction with volatile solvents because enfleurage is a very delicate and lengthy process requiring much experience and labour.

IV. Extraction with volatile solvents

Principle of extraction with volatile solvents is simple. Fresh flowers are charged into specially constructed extractors and extracted systematically at room temperature, with a carefully purified solvent usually petroleum ether. Solvent penetrates flowers and dissolve natural flower perfume together with some waxes and albumins and colouring matter. Solution is subsequently pumped into an evaporator and concentrated at a low temperature. After the solvent is completely driven off in vacuum, concentrated flower oil is obtained. Thus, temperature applied during entire process is kept at a minimum; live steam as in case of distillation, does not exert its action upon delicate constituents of flower oil. Compared with distilled oils, extracted flower oils more truly represent natural perfume as originally present in flowers.

V. Super critical fluid extraction (SCFE)

(Extraction using liquefied gases)

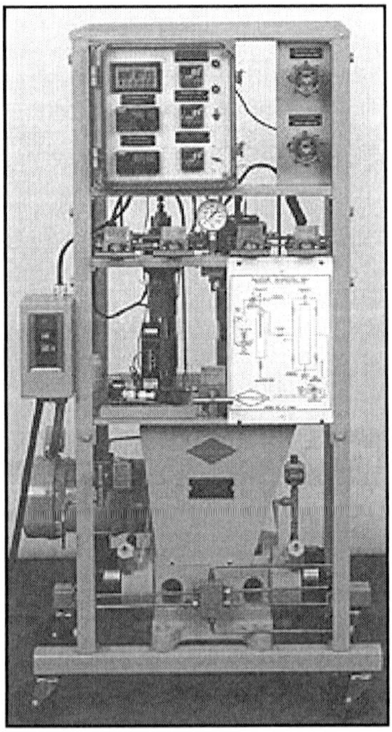

This is emerging as a versatile and important tool to separate components that are susceptible to thermal degradation. It is employed for extraction of flavours, fragrances and perfumes from a wide variety of natural products. This method of extraction is superior and faster than distillation. Higher diffusiveness and lower viscosities of supercritical fluids enable better penetration and faster equilibration. Besides, the solvent power is manipulable, free from surface tension and wetting properties and easily adoptable to isolate highly thermolabile compounds. Carbon dioxide is a favourite solvent by virtue of its cheapness, nontoxicity, noncorrosiveness, non-flammability, easy to handle, needing mild processing conditions during extraction and good solvent power for alcohols, aldehydes, esters and ketones.

Camphor (Page - 63)

Cedar Wood (Page - 66)

Chamomile (Page - 68)

Cinnamon (Page - 71)

Citronella (Java) (Page - 76)

Davana (Page - 82)

Eucalyptus (Page - 84)

Geranium (Page - 88)

Lavender (Page - 92)

Lemongrass (Page - 95)

Linaloe (Page - 105)

Mints (Page - 109)

Ocimums (Page - 114)

Palmarosa (Page - 118)

Patchouli (Page - 123)

Sandalwood (Page - 128)

Vetiver (Page - 131)

Ajowan (Page - 149)

Ambrette (Page - 151)

Caraway (Page - 155)

Celery (Page - 158)

Champak (Page - 161)

Clarysage (Page - 164)

Dill (Page - 167)

Galangal (Page - 171)

Greater Galangal (Page - 174)

Hops (Page - 177)

Jasmine (Page - 181)

Juniper (Page - 186)

Kewda (Page - 189)

Lippia (Page - 192)

Litsea (Page - 194)

Marigolds (Page - 197)

Rose (Page - 201)

Rosemary (Page - 207)

Skimmia (Page - 210)

Sweet Flag (Page - 213)

Sweet Marjoram (Page - 216)

Tea Tree (Page - 219)

Thyme (Page - 223)

Tuberose (Page - 225)

Ylang-Ylang (Page - 229)

In this process, volatility and solubility are employed. Temperature and pressure variations are manipulated to utmost advantage. After extraction, solvent is separated from extract phase by varying operating conditions and is recycled after makeup.

Liquid carbon dioxide is completely miscible with components of essential oils like aldehydes, ketones, esters and alcohols. At same time, proteins, starches, mineral salts and water are insoluble in liquid carbon dioxide. Essential oils obtained by liquid carbon dioxide extraction are superior to that obtained through steam distillation and solvent extraction. Extraction of several natural products such as pyrethrins from chrysanthemum flower, essential oils from anise, caraway, clove, star anise, cinnamon and ginger are increasingly done by this process.

VI. Expression

This method is employed when essential oils are thermosensitive. It is used for isolating essential oils from lemon and orange peels. In general, expression involves squeezing any plant material at great pressures to press out oils or other liquids. The process is carried out by hand-operated presses or crushes in isolated rural areas or by gigantic mechanical presses in industrial centres. In production of citrus juices, oil is unavoidably expressed from fruit. Oil is then separated from juice by centrifuging.

Some technical terms used in essential oil trade

a. Attars

Attars are flower distillates collected over sandalwood oil. Accordingly, there are attars of jasmine, kewda, rose, champaca, khus, maulsari (*Mimussops elengi* L.) kadamba (*Anthocephalus cadamba* Miq.) and mango. Attars are prepared by washing sandalwood oil with one middle and tail distillate from a previous batch to remove lower terpenes from the oil sweetening attar into a mellow and fragrant note. Vapours from water-distillation still are condensed and absorbed in aforesaid washed sandalwood oil. Base oil is then charged to forerunnings of a fresh batch of perfumery plant materials, while tail runnings are employed for preparation of a new batch of base oil. Distillation with fresh charges is repeated till required concentration of natural oil in base oil is obtained. Attars being more stable have longer shelf life. Quantum of flower oil in the product determines quality and price of attar. Attars are used for perfuming tobacco, soaps, agarbathis and hair oils. Alcoholic solutions of attars are now available.

b. Floral and aromatic waters

These are prepared by distilling fragrant materials with water. A specific quantity of fragrant materials is distilled with a specified volume of water for a specific period and a predetermined volume of distillate is collected as per individual formulae of manufacturers. These waters are prepared and marketed as single, double or triple distilled water. These are employed for perfume wafting at sacred places and during social functions. It is also added in traces to cold clean drinking waters.

c. Agarbathis

Agarbathis are the Indian version of Chinese joss sticks and *dhoop* is of incense. Both are burnt slowly to obtain a fragrant smoke. These too have their religious, aesthetic and secular uses. Some industrial houses procure raw agarbathis, add perfumes, pack and market them.

d. Perfumery grade alcohol

This is required as diluent and solvent for production of high class perfumes and flavours, for which olfactory quality of alcohol is a basic and essential requirement. It must be free from higher alcohols and aldehydes. Such alcohols are technically denoted as *Extra Neutral Alcohol* and commercially called *silent spirit,* manufactured from rectified spirit of 90 to 95 % purity obtained from molasses or grains by double or triple fractionation.

e. Flavour encapsulation systems

Flavour encapsulation systems like spray drying, spray cooling, spray chilling, gelatin encapsulation, and cold encapsulation are now increasingly being used by the industry. Spray drying converts a liquid into powder in one step process. Liquid to be dried (feed) is atomized in a stainless steel drying chamber by a nozzle or a rotating disc into a spray which instantly comes into contact with drying air. Due to rapid evaporation, temperature of droplets can be kept far below drying air temperature and is normally 15-20°C below outlet air temperature. Flavour to be encapsulated is specifically formulated to have the highest strength and added to encapsulating medium like gum acacia and modern carriers from modified starches. Carrier choice depends upon emulsification properties, low viscosity in solution at high solid content, flavour retention ability, hygroscopicity, blend flavour compatibility in finished products and stability during storage.

f. Absolute

Absolute is a prepared perfume material. It is a concentrate, entirely alcohol soluble and usually liquid perfume materials. They are obtained by alcohol extraction of concretes or other hydrocarbon types of extracts or from fat extract of plant materials.

g. Concrete

It is a non-purified form of essential oil obtained mostly by means of solvent extraction where essential oil, plant pigments and waxes are present. From concrete, absolute is obtained. Concrete may yield about 45-55% of absolute on a weight basis.

h. Extract

An extract is an alcoholic solution of odorous part of a pomade. This is an intermediate product in preparation of absolutes from pomades.

i. Fixative

In perfumery, a fixative means a material which slows down rate of evaporation of more volatile material in a perfume composition; eg. Sandalwood, patchouli.

j. Oleoresin

An oleoresin is either a natural or prepared material. Natural oleoresins are exudation from tree- trunks and bark. (elm, turpentine, copaiba, balsam). The prepared oleoresins, on other hand are liquid preparations, extracted from plants with help of solvents which can extract oil and resinous matter. They consist entirely or mainly of essential oil and resins.

References

Brogle, H. 1981. Flavour Extraction by High Pressure CO_2. Food F1PP, 3: 19- 21

Caragay, A.B. 1981. Super critical fluids for extraction of flavours and fragrances from natural products. Perfume flavour, 6: 43-48

Guenther, E. 1948. *The Essential Oils*. Vol. I. Van Nostrand Co. Inc., New York, 427p

Payron, L. 1984. Liquid and supercritical carbon dioxide solvent for natural products. Parfums Cosmet Aromes, 55: 47-54

Rajaraman, K. 1984. Extraction of natural products with liquid and supercritical carbon dioxide. Indian Food Ind., 3: 48-51

Stahl, E. and Gerard, D. 1982. High pressure extraction of natural substances with supercritical and liquid gases. Forsch., 176: 1-4

Verghese, J. 1986. *On Essential Oils*. Synthite Industrial Chemicals Private Ltd., Kolenchery, 428p

4

Quality Assurance of Essential Oils

Essential oils are products of secondary metabolism and are secreted by specialized cells located in specified parts of most plants. By virtue of certain characteristic properties, they have wide spread use in perfumery, flavouring and medicine. Essential oils are variable mixtures of terpenoids. Since these oils are produced by plants, composition varies with species and variety of plant besides geography, climate and cultivation practices. Since, chemical components are highly reactive, physical and chemical environment to which oil is subjected to during the processes of its extraction, processing and storage have a great influence on final composition. Entire chain of activities from cultivation, extraction, processing, storage to packing has to be perfectly standardized and regulated to assure end product consistency. Being a valuable item of trade, this is important from the point of view of delivering desired activity as well as safety to consumer.

Physico-chemical characteristics of essential oils

Essential oils are characterized by their physical and chemical properties. They are commonly distinguished by their characteristic odour or fragrance. They possess high refractive indices and most of them exhibit specific optical activity. They are in general immiscible with water and being lighter, most of them float on surface of water. Chemically they are terpenoids and are soluble in organic solvents like petroleum ether, chloroform, ethyl acetate and alcohol in decreasing order. Extent of miscibility with aqueous alcoholic mixtures of given composition is characteristic of an essential oil.

Adulteration of essential oils

Simultaneous with going back to naturals by turn of century, there is an increase in demand for essential oils in world market. Scope for increasing production of natural oils is very much limited. The situation resulted in a tendency for adulteration of essential oils. This has far reaching consequences in aromatherapy and pharmaceutics based on natural oils. Four major types of adulterations are identified.

a. Addition of raw materials miscible with essential oil to increase its bulk

Most common adulteration of this type is addition of cheap vegetable or mineral oils. Most of these materials being not aromatic are usually not detected during odour checking. Since they are non-volatile, they are not detected when analysed by GC-FID either. However, shift in values of optical rotation, refractive index, solubility, acid value and ester value are indicative of such adulterations.

b. Mixing with solvents and synthetic chemicals

There are several cheap synthetic materials like solvents and diluents available in perfumery industry that is used for adulteration of essential oils. They include abitol, benzyl alcohol, benzyl benzoate, diacetone alcohol, dipropylene glycol, isopropyl myristate, dibutylphthalate and diethyl phthalate. The above adulterants are merely diluents and they do not make large changes in odour profile of oil. Addition of these to a level up to 15% may pass un-noticed in normal course of evaluation by organoleptic methods. However, they can be analysed and identified by GC/MS. Classical examples are adulteration of Cinnamon bark oil by adding cinnamon leaf oil, of lavender oil (*Lavandula angustifolia*) by adding cheaper lavandin or spike lavender oil (*Lavandula latifolia*), of patchouli oil by adding gurjun balsam (*Dipterocarpus* spp.) and of peppermint oil (*Mentha piperita*) by adding cornmint oil (*Mentha arvensis*).

c. Upgradation of essential oil of inferior type plants by fortification with active chemicals of synthetic origin

This type of adulteration is most difficult to be identified. Classical examples are upgradation of inferior anise oil by addition of technical grade anethol, basil oil by addition of methyl chavicol and linalool, cassia oil by adding synthetic cinnamic aldehyde, methyl cinnamic aldehyde and coumarin, caraway seed oil by addition of limonene and (+)-carvone, cardamom oil by addition of linalyl acetate, 1,8-cineole and α-terpinyl acetate, dill seed oil by addition of α-phellandrene and limonene, lemongrass oil by adding synthetic citral, *Mentha citrata* oil by addition of linalool and linalyl acetate, palmarosa oil (*Cymbopogon martinii var. motia*) by addition of geraniol, thyme oil by addition of para-cymene and thymol and of vetiver oil by addition of cedrenyl acetate.

Physical parameters of quality control

Each essential oil is characterized by a set of physical properties which can be used individually or in combination to ascertain its purity and genuineness. They are: (a) visual evaluation for colour, clarity, foreign materials etc. (b) organoleptic assessment for odour, appearance, taste, etc (c) specific gravity (d) optical rotation (e) refractive index (f) flash point and (g) solubility in alcohol/aqueous alcohol.

Chemical parameters of quality control

Absolute chemical composition of essential oils can be determined by employing various analysis procedures. They include gravimetry, titrimetry, spectroscopy, spectrometry, thin layer chromatography (TLC), gas-liquid chromatography (GLC), high performance liquid chromatography (HPLC). Chemical parameters like acid value, saponification value and acetyl value are very much indicative of an oil but values fall in a range only. All methods other than chromatography are empirical in nature and results are proximate. A classical example is estimation of citral content of lemongrass oil indirectly by assay of total aldehydes employing reaction of aldehydes with sodium metabisulphite or with hydroxylamine hydrochloride. This method being non-specific is likely to result in amplified values. Conversely, in chromatographic methods, components are physically separated and this is followed by their quantification with help of a detector. The detector is calibrated against known quantity of pure sample of analyte called 'standard'. This method is free from interferences and values are hence absolute. A major drawback is that it is difficult for analyst to possess standards of all analytes due to constraints in availability and high cost of pure chemicals. Further, standards of terpenoids have only short shelf life either. This makes the programme highly expensive. An alternative to this is use of a mass spectrometer in tandem with the GC or HPLC. Chromatographic instrument does the separation and quantification and identification of components is carried out by mass spectrometer.

Storage of essential oils

Essential oils are colourless or lightly coloured and free flowing when they are fresh. On long storage they become darker in colour and highly viscous due to oxidation and resinification. To prevent this, they are stored in a cool and dry place in tightly stoppered amber glass bottles. Exclusion of air by completely filling container with oil prolongs its storage life. Quality of oil is also spoiled during prolonged storage. This deterioration in quality of oil is attributed to a number of chemical reactions such as oxidation, resinification, polymerization, hydrolysis of esters and interaction of functional groups. These processes are activated by heat, presence of oxygen or air and moisture. These reactions are catalyzed by light in some cases and possibly by metals. High terpene containing oils like citrus, pine needle and turpentine are particularly prone to spoilage by oxidation and resinification. Light is less harmful than moisture. Essential oils containing a high percentage of esters, viz, oil of bergamot and lavender turn acidic after improper storage, due to partial hydrolysis of esters. Aldehyde content of lemongrass oil gradually diminishes, yet much more slowly than if isolated aldehyde (citral) is stored as such, perhaps due to presence of some antioxidants. Fatty oils can be preserved by adding antioxidants like hydroquinone. Alcohol containing essential oils like sandal and geranium are quite stable and withstand prolonged storage. Others like patchouli and vetiver improve considerably on aging hence are made to age before use in perfume compounds.

Essential oils should be freed from metallic impurities and moisture followed by its clarification. Then they should be stored in well filled, tightly closed containers at low temperature and protected from light. Bottles of hard and dark coloured glass are well suitable for small quantities of oil. But large quantities are generally stored in aluminium containers or metal drums with tin lining. A stream of carbon dioxide or nitrogen gas blown inside container before it is sealed will replace air above oil and thereby assure added protection against oxidation. Prior to storing, oil should be carefully clarified and any moisture should be removed because presence of moisture is one of main factors in spoilage of an essential oil. Smaller lots can be dehydrated with help of anhydrous sodium sulphate. After addition of anhydrous sodium sulphate, container is shaken thoroughly, kept aside for 24 hours and filtered. Calcium chloride must never be used for dehydration as this forms complex salts with certain alcohols. Large commercial lots are not always easy to clarify. To viscous oil lots like vetiver, sufficient quantity of common salt should be added and mixture stirred for a while and allowed to stand until supernatant oil has become clear which is drawn off. Lower cloudy layer is filtered clear. If filtration through plain filter paper does not give clear oil then kieselghur or specially prepared filtering clay should be used. Care should be exercised in selection of filtering medium as some media like activated carbon may react chemically with certain constituents of oil and affect its quality. Bulk lots of oil may be filtered through filter presses.

Centrifuging at high speed of more than 15,000 rpm is an excellent means of clarifying essential oils. It helps to remove not only moisture but also waxy materials in oil

High phenol containing essential oils like those of cloves and bay, when freshly distilled are in crude form and are dark coloured due to presence of metallic impurities. To get rid of these, sufficient quantity of tartaric acid powder is added to oil lot and stirred to settle the same. Supernatant clear layer is carefully drawn off, while lower layer is filtered until clear. Contents are thoroughly mixed and allowed to separate. Upper oily layer when sufficiently clear is withdrawn off as such. Middle and lower layers are further clarified by filtration. Super centrifuges are of great help here also. If this too fails to eliminate undesired oil colour, oil lot is redistilled or rectified.

Quality control systems around world

A large number of agencies are involved in setting standards for essential oil quality. A few important agencies are:

i. *AFNOR (Association Francàise de Normalisation):* They provide directives and standards for members of European Union states to facilitate intra-community trade. All companies wishing to exchange goods within Europe are therefore obliged to comply with their directives. For essential oils, they provide guidelines and information on various protocols for quality control.

ii. *ISO (International Standard Organization):* Motive of this organisation is to promote development of standardization in areas of intellectual, scientific, technological, and economic activity. For essential oils, they provide guidelines for packaging, conditioning, storage, labeling, sampling and testing. ISO also provides, for a fee, quality standards for individual essential oils.

iii. *ECOCERT International:* This organisation provides organic certification to essential oils after inspection and auditing. Their standards help to differentiate between therapeutic-grade oil and industrial grade oil.

References

Ferren, W. 1991. Instrumental Analysis for Science and Technology. Techno Books, New Delhi.

Guenther, E. 1949. The Essential Oils. Vol. I to VI, D. Van Norstrand Co., Inc. N.Y.

Mahindru, S.N. 1992. Indian Plant Perfumes. Metropolitan Book Co. (P) Ltd., New Delhi.

Verghese, J. 1986. On Essential Oils. Synthite Industrial Chemicals (P) Ltd., Kolencherry, India.

5
Aromatherapy

"The way to health is the way to have an aromatic bath and a scented massage everyday" **Hippocrates**

Aromatherapy is the art and science of using fragrant volatile plant oils and herbs including essential oils, for psychological and physical well-being of man by promoting natural healing and health. It is an emerging art of healing through essential oil, which has got therapeutic effect on body, mind and soul. Essential oils give calmness, emotional and hormonal balance, stress relief and rejuvenation. Medicinal and spiritual properties of scented herbs were noted by ancient Egyptians and Babylonians. Embalming of body of Egyptian Pharaohs was a practice to purify body for afterlife, whereas Egyptian priests used oils and incense as healers. Books were written in ancient Arabia indicating uses and benefits of certain aromas. Perfumed mortar was used by Babylonians to build temples and mosques, for premises cleaning and cleansing aromas. Ancient Greek believed sweet smells were of divine origin and they used to take aromatic bath to drench them in purity. Aromatic bath and scented massage were ways to health for Romans.

The term "aromatherapy" is attributed to a French cosmetic chemist named Rene Maurice Gattefosse. There is a story behind his working on aromatherapy. As he worked in his laboratory in early 1920, he severely burned himself. Without knowing effect, he plunged his arm into lavender essential oil. To his wonder, burns healed rapidly, with little scarring. After this incident, Gattefosse dedicated remainder of his life for aromatherapy or healing power of scented healing oils. Modern research has then reinstated the therapeutic and healing properties of essential oils. Aromatherapy restores or enhances mental, emotional, physical or spiritual health. Incorporating aromatherapy into life enhances overall health, beauty and psychological well-being. It reduces stress, improves sleep and gives more energy. Essential oils give plants their characteristic odours. Since, aromatic essential oils are used, therapy involving them is named as aromatherapy.

Plants, which are storehouses of cosmic energy are the main source of energy for mankind. In this way, human consciousness can be regarded as a plant living within us that grows in accordance with our perception and consciousness of life. Cosmic energy is to be found in every part of a plant. In aromatic species, this information is concentrated in certain volatile, fragrant substances with an oily consistency known as essential oils. Essential oils are the heart of aromatherapy. Essential oils are formed with help of solar energy acting on aromatic plant's secretory cells. Plants keep it in tiny glandular pockets which burst open when they are given stresses.

René Maurice Gattefossé in 1920's described practice of using essential oils taken from plants, flowers, roots and seeds in healing. It is the chemical properties of essential oil that gives therapeutic value. Vapours are used in some cases of aromatherapy, whereas mostly oil is rubbed on to skin or ingested in a tea or other liquid. Some aroma therapists even consider cooking with herbs a type of aromatherapy.

Many essential oils are used for their smell, their antibacterial, anti-fungal or anti-inflammatory properties. Lavender oil is a classic example in this line used for burns and scent of it reduces depression and anxiety. Matthiole, a 16th century botanist, prescribed that lavender is a panacea which can cure epilepsy, apoplexy and mental problems. The major constituents of lavender are alcohols such as borneol, geraniol and linalool. It also contains a high percentage of phenol, a strong antiseptic and antibiotic. According to Matthiole, lavender oil is effective for burns, cystitis, vaginitis, and leucorrhea. As a herbal tea, lavender is also good as a morning tonic for convalescents, as a digestive after meals, for rheumatic conditions and at first appearance of a cold or flu. To prevent varicose veins, Ryman advises to massage legs with oil consisting of 3 drops cypress oil, 2 drops each of lavender and lemon oil, and one ounce of soy oil.

Healing power of tea tree oil is an anecdote. Daniele Ryman quotes his own experience with tea tree oil as a remedy for swelling and throbbing due to rose thorn injury. Tea tree oil is a time honored aromatherapy remedy for ringworm, athlete's foot and other fungal infections. It is also effective against many bacteria present in common infections, including some staphylococci and streptococci. Rose oil is prescribed for frigidity due to its aphrodisiac properties. Rose oil is a good tonic for women who are suffering from depression. Rosemary can be used to treat arthritis and muscle pain and is a stimulant, when used in morning bath. A variety of oils like lavender, thyme, chamomile, rosemary, lemon, clove, tea tree, geranium, peppermint, eucalyptus, rose and sandalwood are used in aromatherapy.

There are two main ways to use fragrance in healing. One is through inhalation, which has its significant impact on mood and emotion, but also produces physical reactions such as lower blood pleasure. The other route is physical application of essential oil to body by massage. By using aromatherapy oil, the fragrance is also inhaled. It is known that merely smelling a fragrance can influence us physically and emotionally by altering hormone production, brain chemistry, stress levels and general metabolism.

Effect of smell

Smells invoke long-term memory and make the past present as none of the other senses can. Smell bypasses thalamus in brain and penetrates directly to "olfactory brain", where it produces pleasure or repugnance. Smell is our most direct means of communication with nature. We smell every breath we take, constantly monitoring world around us although we are not always conscious that we are doing so. An involuntary reaction is created by every scent in human minds. To fully understand aromatherapy, knowledge on two physiological processes viz. how the olfactory apparatus works and how essential oils are absorbed into the body is a must.

Olfactory function

Sense of smell is related to sense of taste. Olfactory sense receptors are in receptor nerve endings in direct contact with outside as well as brain. Certain brain waves called "contingent negative variation," are very sensitive to emotional changes and are activated by particular fragrances. Different odours stimulate different brain centers and induce neurochemicals that affect us in a number of ways. For example, "euphoric" odors stimulate thalamus to secret enkephalins, the natural painkillers. Aphrodisiac scents jasmine and ylang-ylang stimulate endorphin producing pituitary gland. Sleep is induced by sedative odors like marjoram which stimulate secretion of neurochemical serotonin.

Essential oil absorption through skin

Absorption of essential oil through skin is quick and easy due to their lipid solubility. Essential oils enter blood stream through small capillaries. Nervous and lymphatic systems are activated by presence of aromatic oils. Studies show that after a full body massage with 2% dilution of lavender essential oil in vegetable oil, detectable amounts of linalool and linalyl acetate are found in blood. Similarly garlic rubbed on feet can later be smelled on breath. It shows that massage with fatty oils is also absorbed by skin.

Individual can be identified by linking of a particular odour as our finger prints, influenced by diet, gender, heredity, health, medication, occupation, emotional state and mood. This linking influences choosing of friends and lovers.

Safety aspects to use essential oils

Compounds in the oil, dosage, frequency used and method of application are to be understood before using essential oils. Because essential oils are concentrated, here are a few guidelines to ensure safe and effective use.

- Use only pure essential oils
- Must be diluted in a proper base before internal or external use. If used concentrated, they can cause burning, skin irritation and photosensitivity except in lavender oil or tea tree oil used on burns, insect bites, wounds, eczema, certain types of cramps, headaches, neuralgia and rheumatism. Before application, test sensitivity by putting one drop on inside of wrist and wait for any irritation within 24 hours
- For making essential oils miscible with water, a safe natural emulsifier may be used. A concentration of 70 to 95% alcohol, vegetable oils, honey and yogurt are used as emulsifiers and diluents.
- Oils such as allspice, cinnamon and clove which cause irritation to mucous membrane and skin should be used carefully. Eye contact should be always avoided.
- In uncertain cases, patch test with a 2% dilution in crook of arm or on back of neck at hair line can be done. See for Twelve hours for reactions such as redness or itching. In case, reaction develops, try a less potent dilution, or use an appropriate substitute.
- Photosensitivity is another type of reaction, which causes uneven pigmentation of skin upon exposure to sun light. Citrus oils like bergamot, bitter orange, lime, lemon, grape fruit angelica, cumin and verbena cause such reactions. In case of use of these oils on skin, do so at night, stay indoors, or wait at least four hours before exposing skin to ultraviolet light.
- Essential oils should be kept out of reach of children. Use only one third to one half the adult dosage for children and select only nontoxic oils like lavender, tangerine, mandarin, neroli or roman chamomile.
- Avoid using essential oils during first three months of pregnancy, especially to those women who are prone to miscarriage. The safe essential oils for this time are gentle floral oils such as rose, neroli, ylang-ylang, chamomile, jasmine, geranium, sandal wood and spearmint.
- Use essential oils cautiously with those who are elderly, convalescing, or have serious health problems such as asthma, epilepsy or heart diseases.
- Continuous use of a particular blend is not advisable since it may affect kidney and liver functions. After using a particular blend for about two

weeks, change to another especially when whole body is treated. Facial blends or treating a small part of body with low dosage is not found harmful.

- Essential oils as such should not be consumed internally for therapeutic purposes.
- Over exposure to essential oils through inhalation or skin application may result in nausea, headache, skin irritation, and emotional unease. For skin irritation and eye contamination, dilute with straight vegetable oils but not water for relief.
- Professional aromatherapy massage treatments may give relaxation, but not cure since it is without diagnosing disorders. Do not massage strenuously near varicose veins, bruises or other types of broken veins.

Dilution of essential oil for therapy

The most effective way to dilute essential oils is in carrier oil. High quality vegetable oils such as almond, apricot, hazelnut, olive, grape seed, coconut or sesame are good carriers. A safe and effective dilution for most aromatherapy applications is 2% for adults. Vegetable oils, which are fixed oils, contain vitamins A, E and F which give soothing, skin-softening and nourishing effect and are also rich in nutrients that enrich skin.

Use of emulsifiers

Emulsifiers bind water and oil together so that they will not separate. The most common natural emulsifiers are:

a) Beeswax

Beeswax is the most common natural emulsifier for home made cosmetics, the best for holding water and oil in suspension. Depending on concentration of bees wax, thickening of a lotion or hardening of a lip balm occurs. If beeswax is brittle, it is probably old. If it is dark, it may contain propolis, an antibacterial material used by bees to seal their hives. A bit of propolis in beeswax is good for cosmetics.

b) Lanolin

This is the oil that is removed from sheep's wool. In structure, it is much like oil of our own skin glands and hence easily adsorbed. Thick anhydrous lanolin is the least desirable, because it does not mix well with water. Hydrous lanolin containing a little water is used in making lotions. Liquid lanolin is a lotion on its own and it can also be used as an ingredient for lotions and creams.

c) Glycerin

Glycerin is a clear sweet sticky product derived from plants or animals or as a byproduct of soap making.

d) Lecithin

This is derived from soybeans and also found in egg yolk, once commonly used to emulsify bath and hair products. It increases spreading property of a product and leaves skin feeling very smooth.

Storage and shelf life

Store essential oils and carrier oils away from heat and light to preserve their freshness and potency. When stored properly they have a shelf life of several years. Refrigeration of all vegetable oils is highly recommended. Citrus oils have the shortest shelf life of one year. Oils of frankincense, myrrh, sandal, vetiver and patchouli last long and improve on aging.

Treatments with essential oils

Essential oils are extremely concentrated. Most of them are at least 50 times more potent than herbs from which they are extracted. One drop of essential oil often represents potency of one ounce of plant material. This indicates its healing potential and potential hazards if used improperly.

a. Heart and circulation

Lymph nodes located in throat, groin, breasts and under arms act as centers for filtering blood. For lymphatic massage, true bay (*Laurus nobilis*) and lemon oil are good with calendula as carrier oil. Basil, rosemary, thyme, marjoram and clove improve general circulation.

b. Digestive system

Brain stimulates whole digestive system on receiving impulse of a food aroma. As an immediate reaction digestive fluids are released in mouth, stomach and small intestine. Essential oils such as rosemary, basil, cumin, anise, coriander, ginger and cinnamon are used in culinary and they also help digestion.

c. Bowel problems

Bowels can become irritated and infected by various foods. Even excitement or stress can agitate bowels. Ginger, peppermint, fennel, coriander and dill help counter gas. Peppermint is specific for irritable- bowel syndrome. For constipation, use rosemary or black pepper. For diarrhea use cypress, cinnamon and myrrh. Garlic is one of the best ways to eliminate worms for whole family including pets. Rosemary, thyme, tea tree and chamomile kill many types of worms; chamomile also decreases resulting intestinal inflammation.

d. Respiratory system

Ninety per cent of respiratory ailments are caused by viruses. Oils of thyme, rosemary, peppermint, tea tree, eucalyptus, bergamot, black pepper, Melissa and hyssop inhibit most flu viruses. Lemon and eucalyptus oils are effective against bacteria that cause staph, strep and pneumonia infections. A 2% dilution makes an effective antiseptic gargle or vapor steam. Steam treatment carries essential oils directly to sinuses and lungs, and provides warm, moist air to help open nasal and bronchial passages.

e. Musculo-skeletal system

Bones and muscles give form to body and permit physical movement. With degenerative conditions such as arthritis and rheumatism, entire body must be treated especially digestive and eliminative systems. Use anti-inflammatory essential oils such as grapefruit, juniper and helichrysum to stimulate circulation and eliminate toxins.

f. Nervous system

Nervous system provides intricate connection between mind and body. As a result of mental or emotional responses, a problem in one area of body may affect another. For stress in general, bergamot, chamomile, lavender, melissa, clarysage, neroli, rose or jasmine are good. For insomnia due to mental agitation or overwork, clarysage, marjoram, ylang-ylang and neroli can work well.

g. Urinary tract

Urinary system, consisting of kidneys and bladder, regulates body's water content and salt balance and eliminates waste. Antiseptic diuretics to treat bladder infections include cedar wood, tea tree, bergamot and fennel.

Aromatherapy massage

In an aromatic oil massage, blood takes oil to where it is needed. For any type of massage, oils are only lightly fragranced since a strong smell will overwhelm giver and taker. Massage parlor should be a well-ventilated room. Air out the room between clients by opening windows or using a fan or other wise, use an air filter to remove scent from air.

Lightly misting a floral water or hydrosol over person before receiving massage gives a special touch. To relax the person, place a warm herbal compress over tight muscles. Simply add a few drops of any of essential oils to very warm water, submerge a soft cloth, swish it around, wring it out and place it on skin for two to five minutes or longer and repeat it as time permits. For an extended treatment, place a towel or a heating pad over compress. After removing compress, pat skin dry with a warm towel. A compress over eyes, both before or after a massage, is very relaxing and relieves headache. Use only gentle oils for this.

Special types of massages

For acupressure a small amount of oil can be applied to fingertips. A few drops of essential oil will fill the room with fragrance. Power of penetration, diffusion and volatility make it possible for essential oils to get assimilated by respiratory tract and lungs through skin. The blood carries them to organs, and gets distributed throughout entire body. Many neuropaths consider external use as best method of administering essential oils, whereby a pleasant experience can be combined with useful treatment.

a. Facial massage

More muscles are concentrated in face than anywhere else in body. Facial tension contributes to wrinkles and to poor blood circulation in face. A good facial massage takes much less time than a body massage. Always use very lightly scented oils around face and be sure to keep oil well away from eyes. Often a thick facial cream works the best. A good sequence for a facial massage is to begin at chin and work up, against gravity. If there's time, give tight shoulders and neck a quick rub. Be sure to use gentle, light strokes on face since kneading face too much only contributes to more wrinkling. A light tapping motion with fingers as if you are typing, gives a stimulating massage. Give massage to the ears and jaws also.

b. Spot massage and liniments

Liniments reduce muscle and joint pain when rubbed into skin, but contain a higher concentration of warming essential oils, which are potentially irritating. Liniments are designed for spot massage on particularly painful areas only.

Different health and body care therapies

Condition of our hair and skin reflect our inner health and beauty. Nature's gifts of herbs and essential oils offer many benefits for body care.

a. Inhalation

This is a simple and efficient treatment method for colds, bronchitis, sinusitis, etc. Put 2-3 drops of each essence from sage, pine and lavender in a bowl of hot, but not boiling water. Completely cover head with a towel and inhale steam for at least 10-15 minutes. Repeat this two or three times a day.

b. Fragrant mixtures for handkerchief or pillow

Putting a few drops of essential oils of eucalyptus, thyme, and lavender on a hand kerchief or on pillow and deeply breathing it give relief from cold, cough and flu.

c. Aromatic bath

Healing baths are an important aspect of treatment with natural cures, both in the form of complete baths or as hand and foot baths. Nerve endings of entire body are concentrated in skin of hand and feet. Essential oils can be added to water in undiluted drops. Since, oils are hydrophobic and lipophilic, oily skin easily absorbs undiluted oil much more quickly than those diluted in a vegetable-oil base. Hot water also causes your skin to be especially receptive to absorbing essential oils. Non-irritating oils such as lavender, tea tree and geranium, 15 drops (use about half of this amount for hand and foot baths) are safe and delightful. Fill tub with water and add oils just before bath and agitate water well to distribute oils. If any skin irritation is felt, get out of tub, rinse with cool water and apply straight vegetable oil to skin.

Steam bath encourages sweating, which aids circulation and helps to flush out the system and revitalize skin. This is adopted by different cultures of people for treating of diseases and as part of spiritual practice.

d. Aromatic body powders

Arrowroot, cornstarch and white clay make good bases for natural aromatic body powders for babies or adults.

e. Aromatic hair care

Beautiful, shiny and vibrant hair reflects self-image. Hormonal fluctuations, diet, lifestyle and stress influence appearance and health of hair. Pollution, harsh detergents, chlorine, blow drying and excessive sun exposure adversely affect vitality of hair. Acid shampoos and vinegar alter electrical charge of hair, reducing its tendency to become "flyaway". They also remove soap and detergent residues, leaving hair shiny and soft. Add 1-3 drops of essential oils to normal shampoo or to a mild baby shampoo to make it fragrant. For different types of hairs, different treatments are required. Lavender and rosemary oils are good for normal hair. For dry hair, use mild shampoos containing fatty acids and moisturizers. Hot oil treatments are specific for dry hair, dry scalp and dandruff. Oily hair and skin are caused by excess sebum production. Oils come from scalp, so hair is much oilier near roots than at tips. To remove excess sebum and keep oily hair bright requires frequent washing with a mild shampoo. Herbal oil preparation can be used to treat dandruff, lice and falling hair. A scalp massage with jojoba oil, vitamin E and essential oils such as rosemary and *Aloe vera* helps to keep hair roots healthy and promote hair growth.

Facial care

In aromatherapy facial treatment, basic techniques are cleansing, steaming, exfoliation, masking, toning and moisturizing. Use of these methods and choice of essential oils and herbs, should be determined by skin type.

- For *oily* skin: Essential oils of lemon and lavender or mint or sage or carrot.
- For *dry, devitalized* skin: Essential oils of rosemary and lemon, or Melissa, or verbena.
- For *acne and blackheads*: Essential oils of lavender and juniper, or sage or chamomile, or geranium. To cleanse and care of skin and also to *prevent wrinkles*: Essential oils of lemon and carrot, or chamomile, or orange, or patchouli are to be used.

Cleansers

Healthy skin produces lactic acid and eventually regains its acid mantle. Sodium lauryl sulphate is the basis of most liquid facial and body soaps, as well as shampoos. Ground oat meal is good for washing face.

Steaming

Steaming helps soften sebum and unclog pores. Steam carries volatile oils directly to face, so essential oils or fragrant herbs can be added to steam water.

Exfoliation

Exfoliation is removal of dead skin cells from epidermis. Done properly, this brings young, fresh skin to surface and stimulates cell growth in lower layers. Exfoliation also gives impression of erasing wrinkles. Alpha hydroxyl acids encourage exfoliation by loosening tight bond that holds surface skin together.

Masks

Masks can absorb moisture, feed or remineralize the skin. Several clays may be used as a base: white and green clays are for any skin type; red or yellow clays are very absorptive for oily skin; and gentle blue clay is for sensitive skin. Colours are natural and reflect minerals found in clay. Eggs, fresh fruits and vegetables, oats, cream of wheat, yogurt and nutritional yeast are just a few of other possibilities for mask ingredients. Fruits and herbs containing skin-softening enzymes act as exfoliants. Essential oils or ground herbs can be added to increase the skin-healing properties of mask. To make a mask, mash ingredients into a thick paste, adding hydrosols, herb tea or aloe to moisten dry ingredients. Avoid sensitive areas around eyes and edges of mouth. Facial masks can be left on from 5-20 minutes, don't allow mask to dry or pull so much that it becomes irritating. Wash mask off with warm water, apply massage oil, if desired and gently pat skin dry.

Toners

They increase circulation, improve skin tone, and reduce wrinkles and enlarged pores. Toners can also serve as moisturizers and offer a good alternative to oil-based moisturizers for those with very oily or problem skin. Facial toners can be made with plain apple cider, wine or other vinegars.

Aromatic hydrosols

An aromatic hydrosol produced during essential oil distillation can be a valuable addition to beauty regimen. They can be used alone as toners, or added to other toners ingredients, such as *Aloe vera*.

Fragrant waters

Fragrant waters are made by adding essential oils to distilled water. We call them "fragrant waters" to distinguish them from true aromatic hydrosols. Adding *Aloe vera* juice to fragrant water after daily shower cools down on a hot day.

Moisturisers

Facial oils are made of essential oil diluted to 2-3 % in carrier oil. They do not contain the water that creams and lotions provide, but when applied directly after a toner is effective skin treatments. Liposomes can also be added to facial oils, but because the watery liposomes separate from oil, the mixture should be shaken before use. Apply moisturizers in a thin layer over entire face and neck while skin is still damp with toner to help seal in precious moisture.

Skin care properties of herbs and essential oils

Here are just a few effects of essential oils upon the skin:

- Penetrate dermal layer of skin where new cells are developing;

- Stimulate and regenerate; produce healthy skin cells quickly following sun damage, burns, and wrinkles or healing of wounds.
- Reduce bacterial and fungal infections, and other related skin problems.
- Soothe delicate, sensitive, inflamed skin.
- Regulate sebaceous secretions, balancing over-or under active skin;
- Promote release and removal of metabolic waste products.
- Contain plant hormones that help balance and alleviate hormonally related skin problems.
- Affect mental and emotional state positively, thus alleviating stress-related skin problems.

Aromatic miscellany

Gentle shaping and moisturizing encourage healthy growth and strengthen nails. Herbal-tea soaks or herb-infused oil treatments of comfrey, oat straw and horsetail can strengthen nails and cuticles. Lip balms protect lips from drying wind and cold conditions.

A good rub with a few drops of essential oils of rosemary, pine or lemon on one area of body or whole body stimulates refreshes or calms very effectively. Fragrances of citronella, lemon grass, eucalyptus, geranium, and mint are used to keep away mosquitoes and flies. Natural aerosols with ionized oxygen in countryside and mountain areas with electrically charged air and enriched with volatile aromas has a bacterial and stimulating effect on respiration and entire organism.

References

Balz, R. 1999. *The Healing Power of Essential Oils.* Motilal Banarsidass Publishers Pvt. Ltd., 203p

Guenther, E. 1948. *The Essential Oils.* Vol. I. Van Nostrand Co. Inc., New York, 427p

Guenther, E. 1949. *The Essential Oils.* Vol. II. Van Nostrand Co. Inc., New York, 852p

Guenther, E. 1949. *The Essential Oils.* Vol. III. Van Nostrand Co. Inc., New York, 777p

Guenther, E. 1950. *The Essential Oils.* Vol. IV. Van Nostrand Co. Inc., New York, 752p

Junimann, M. 1999. *Enchanting Scents.* Sri Satguru Publication, Delhi, 123p

Keville, K. and Green, M.1999. *Aromatherapy-A Complete Guide to the Healing Art.* Alternative Medicine Series No.1. Sri Satguru Publications, Delhi, 56p

6

Major Sources of Aromatic Oils

Camphor

Scientific name: *Cinnamomum camphora* Nees & Eberm;
Syn: *Laurus camphora* (Linn.) *Camphora officinarum* Nees
Family: Lauraceae
Vernacular names: *Sanskrit:* Karpurah; *Hindi:* Karpura; *Malayalam:* Karpuramaram; *Tamil:* Karpooram; *Kannada*: Karpooram; *Telugu:* Karpuramu

Camphor plant is valued for presence of camphor in its wood and leaves. Camphor is chiefly a pharmaceutical product though it is used in preparation of artificial essential oils like lavender and lavendin. It is used as incense and in balms of various kinds. It is employed as a masking agent in perfumery and sometimes used to achieve lift in perfume blends. Redistilled brown oil is used directly in soap perfumes for a masking effect. Oil is a source of safrole which is a starting material for production of various perfumery chemicals. In 13th century, Marco Polo noted that camphor oil was highly valued by Chinese as a medicine, scent, and embalming fluid. Plant contains a volatile oil comprising camphor, safrole, eugenol and terpineol. It also contains lignans. Camphor is irritant and antiseptic; safrole is thought to be carcinogenic. A white crystalline substance derived from stems, roots and other parts of tree, also called camphor has powerful antiseptic, circulatory stimulant and calming effect in cases of hysteria, neuralgia and general nervousness. Oil obtained by distillation is used to treat diarrhoea, rheumatism and muscular pains. It is very useful in bronchitis and pneumonia. Oil also stimulates uterus, menstruation and uterine hemorrhages. Since, oil contains tannin, it has an astringent effect and increases constipation unless mixed with laxative. It may also be applied to skin problems, such as cold sores and chilblains.

Habitat and distribution

Camphor tree is a native of Japan, China and Taiwan. It is distributed over tropical and sub-tropical countries of Asia, Africa and South America for its woods from which camphor oil is derived. In India, plant is cultivated in Dehra Dun, Calcutta, Bangalore and Nilgiri hills.

The plant

Camphor tree is a large handsome evergreen tree with an enlarged base and reaches a height of 8-18 m, produces red leaves that turn dark green as they mature and are opposite or alternate, usually triple nerved, leathery, shiny whitish beneath, up to 5-7 cm long. Leaves and twigs of these plants have a strong camphor smell. Leaves are shed every year during February-March, but simultaneously new leaves appear. Plant produces small fragrant yellowish flowers and oval red berries. Flowers are bisexual, borne on axillaries, lateral or sub lateral cymes or panicles; perianth tube short, funnel shaped, enlarged in fruit, lobes 6, sub equal, persistent and partly truncate; perfect stamens 9 or fewer with introse 4-celled anthers; ovary sessile, at bottom of perianth tube narrowed into thick style; fruit seated on enlarged perianth tube.

Camphor is formed in oil cells distributed in all parts of tree. These cells begin to form early in growth of plant and are filled with a yellow oil from which camphor is slowly deposited. Oil becomes colourless progressively and volatile and irregular masses of camphor appear. Formation of camphor is brought about by an enzyme present in growing parts of tree, particularly in tissue within cambium region. Each layer of wood, as it is formed, is enriched by camphor.

Cultivation

Camphor tree grows the best at an altitude of 1350-1500 m with temperature not going below 33°C. In Nilgiris it does well up to 2100 m above MSL. Plant can be successfully cultivated where rainfall is up to 1000 mm/annum. Fertile, well drained sandy loam soils are best suited for cultivation of camphor tree. Deeply tilled clayey soils are also suitable, provided rendered porous by mixing leaf mould and sand supplemented with artificial fertilizers.

Propagation is chiefly through seeds and rarely through layers, branches, cuttings, root cuttings and root suckers. Fresh ripe fruits are collected either direct from tree or soon after they fall. Removal of pulpy seed coat and presoaking of seeds in water for 24 hours enhances seed germination. Seeds are sown in nursery at a spacing of 6-8 cm in rows of 25-30 cm apart and irrigated regularly. Seeds start germinating after 3 months of sowing. Nursery is maintained weed free. After 12-16 months seedlings are transplanted in main field in 60 cm cube pits, 2-3 m apart. Application of organic manures and inorganic fertilizers has proved beneficial. Plants are trimmed to a height of 1.5-2 m and maintained as bushes to facilitate picking of leaves. Leaf blight disease in camphor is caused by *Glomerella singulata* which can be controlled by spraying difolatan and benlate. Leaves and twigs are harvested every year and distilled to produce camphor oil. Wood over 50 years of age is also used for distillation. Bushes are harvested 3-4 times a year.

Extraction and utilisation

Camphor oil is obtained by distilling leaves, twigs and wood. Crude camphor oil is separated in various fractions as white, brown and blue camphor oils. White camphor oil is generally not used as such in perfumes, but it serves as a starting material for production of a number of perfumery chemicals such as cineole, terpineol, menthol and thymol. There is little difference in total yield of camphor when two or four pickings are taken in a year. Tender leaves and plants grown in open contain more camphor. Yield of camphor and camphor oil is 50-80 kg/ha which varies widely with part used and geographical location.

Camphor tree exhibits wide variability in different locations. Samples from Calcutta (India) recorded following properties.

Property	Leaves	Branches	Trunk
Colour	Colourless	Light brown	Light brown
Optical rotation	30°6'	26°21'	28°48'
Acid value	2.238	2.980	3.125
Ester value	12.148	3.965	4.396
Solubility	1:1	1:2	1:2
	(90% alcohol)	(80% alcohol)	(80% alcohol)

Essential oil distilled from branches, wood and root is obtained as a semi solid mass. Yield is generally 1-1.2% and wide variability is reported with locations.

Leaf oil has following constituents viz. sabinene 1.47%, α-phellandrene 0.17%, α-terpinene 0.24%, terpinolene 0.30%, furfural 0.16%, piperitone 2.4%, geranyl acetate 0.22%, cuminaldehyde 0.15%, safrole 13.4%, eugenol 0.12%, cinnamyl alcohol 0.18% and traces of more than twenty other compounds.

Cedar Wood (Himalayan)

Scientific name: *Cedrus deodora (Roxb.)* Loud.; Syn: *Pinus deodora*

Family: Pinaceae

Vernacular names: *Sanskrit:* Snehavriksha, Devadaru; *Hindi:* Deodar; *Bengali:* Toon; *Malayalam:* Devadaru; *Tamil:* Toon-maram; *Kannada*: Devadari; *Telugu:* Devadari; *Punjabi:* Pahari-keli

Cedar wood oil, also called Himalayan cedar wood oil, is obtained from wood chips, shavings & saw dust of this gymnospermous plant. Oil is used with other essential oils as a fixative and diluent in soap perfumes, in sanitary supplies and polishes and for masking odours in many other industrial products. Special grades of this are used for oil immersion lenses and as a tissue clearing agent in plant and animal histological work. Oil is biologically active against mosquito (*Anopheles stephani*). A low concentration of cedar wood oil (0.445%) proved sufficient to knock down 50% of mosquitoes under laboratory conditions.

Cedar is also one of the most valued Indian timbers. Primary use of this is for construction works, as beams, floor, boards, posts, door and window frames. Wood is fragrant, insect repellent, rot resistant and quite durable. Wood also possesses diaphoretic, diuretic and carminative properties. Bark is astringent and useful for treating fevers, diarrhea and dysentery. Oleoresin of deodar and dark-coloured oil obtained from wood cure ulcers and skin diseases.

Habitat and distribution

Deodar forests are common from Kashmir (Kishanganga, Kestwar and Jhelum valley) to Garhwal (Uttaranchal). Tree is found growing wild in forests of Jammu and Kashmir, Himachal Pradesh and Uttaranchal. It occurs at altitudes of 1200-3000 m, but growth is more luxuriant between 1800 and 2600 m.

The plant

Other species yielding cedar wood oil are belonging to genus *Juniperus, Cedrus* and *Cupressus*. *Juniperus virginiana* L. yields American red cedar wood oil or Virginian cedar wood oil, *J. procera* Hochst. East Indian cedar wood oil, *J. mexicana* Spring. Tenean cedar wood oil, *Cedrus atlantica* Manatti. Atlas cedar wood oil and *Cupressus fenebris* Endl. Chinese cedar wood oil. The oils obtained from these sources differ chemically and olfactorily.

Cedrus deodora is a tall, evergreen, graceful tree attaining a height up to 100 m and 2.4 to 3.6 m in girth. Growing shoots and branches are drooping. Tree has typical cone-like appearance, but as it grows, it becomes rounded or flat with spreading horizontal branches. Bark of cedar wood tree is greyish-brown with vertical or diagonal cracks.

Tree has a tap root system. Leaves are usually dark green, sometimes bluish green 2.5 to 5 cm long.

Extraction and utilisation

Oil is obtained from sawdust, wood shavings, stumps and roots of cedar wood by steam distillation. Oil content from saw dust ranges from 2 to 4.5%. Steam distillation at 2.38 kg/cm^2 steam pressure gave an oil yield of 4.41% with 33.6% carbonyl content. Oil can be obtained economically if steam is injected at 4.5 kg/minute in stills for 14 hours. Longer durations added to cost of production.

Petroleum ether having boiling point 70-80°C yields oil to extent of 18.6%. Alcohol gives better yield (21.7%) but product obtained is resinous.

Physico-chemical properties of oil

Specific gravity (15°C)	0.9530 - 0.9756
Optical rotation	+34° to 53°8′
Refractive index (20°C)	1.515 - 1.523
Acid value	1.6 - 5.6
Ester value	4.9 - 20.5
Ester value after acetylation	30.8 - 39.2

The constitutents of oil identified by GCMS are α-pinene 0.08%, limonene 1.20%, *p*-methyl-Δ^3– tetrahydroacetophenone 0.60%, *p*-methyl acetophenone 0.34%, β-caryophyllene 0.34%, δ-cadinene 0.69%, β-cedrene 1.40%, α-cedrene 15.82%, β-hemachalene 12.34%, α-hemachalene 30.83%, sesquiterpene alcohols 2.26%, deodarone 5.38%, cedrol 1.36%, hemachalol 1.33%, sesquiterpene ketones 5.61%, cis-atlantone 2.43% and trans–atlantone 6.52%.

Chamomile

Scientific name: *Chamomilla recutita;*
Syn: Matricaria recutita, M.chamomilla Linn.
Family: Compositae (Asteraceae)
Vernacular names: *Hindi*: Babuni ki phool, Babunah; *Bengali*: Babunphul; *Tamil*: Seemaseventhi poo; *Gujarati*: Babuna; *Marathi*: Babuna; *Punjabi*: Babuna; *Malayalam*: Shima jevanthi pushpam; *Kannada*: Shime- shavantige; *Telegu*: Sinacha mauli pushapamu.

Diverse products such as essential oil, infusions, tinctures and fluid extracts are prepared from this plant. Its essential oil is used in high class perfumes to a limited extent. It is also used as flavouring agent in various products such as alcoholic and non-alcoholic beverages, herbal tea, ice-creams, ice candy, baked foods and chewing gums.

Oil has many therapeutic uses. It is antimicrobial, antispasmodic, antipyretic, anti- inflammatory, expectorant, carminative, anthelmintic, analgesic, sedative and diuretic. It promotes wound healing and acts also as a tonic. Chamazulene produced from proazulene on distillation, is highly antiallergenic. Hence, oil can be externally applied to sore, itchy skin, sore nipples and eczema. It is used in infant ailments such as teething troubles and stomach disorders. Herb is valuable for pain, indigestion, acidity, gastritis, peptic ulcer, Crohn's disease and irritable bowel syndrome and also in hay fever and asthma. Infusion is a mild sedative and digestive and used as a common beverage in Europe. One of the key constituents spiroether is a strong anti- spasmodic and eases tense muscles and menstrual pain. Dried herb and tinctures are used in bitter tonic beverages and elixirs. It is also used for nervous tension and irritability. It also relieves eye strain. Flower odour induces sleep and drives away noxious insects.

Habitat and Distribution

German chamomile is indigenous to Europe. It is grown in Germany, Hungary, France, Russia, Yugoslavia and Brazil. It was introduced to India during Mughal period; now it is grown in Punjab, Uttar Pradesh, Maharashtra and Jammu and Kashmir.

The plant

German or Hungarian chamomile (*Matricaria chamomilla* L.) is the most widely and commonly cultivated species. It is a much branched, erect spreading annual herb, 60-90 cm tall and glabrous, aromatic, bitter taste. Leaves are pinnate with narrow and linear leaflets. Flowers are solitary, 1.3-2.5 cm in diameter with flowers borne on hemispherical or conical hollow receptacles, surrounded by involucre of 2-3 rows of small imbricate bracts. Ray florets are around 10-20 in number, whitish or yellowish, later becoming reflexed and disc florets are numerous, yellow and tubular. Peduncles are 2.5 cm long, dark brown or greyish yellow and achenes are with 3-5 faint ribs.

Other related species are:

Anthemis nobilis L. Syn. *Chamimelum nobile* known as Roman Chamomile

Ormenis multicaulis Braun-Blanquet & Maire is termed as Moroccan Chamomile.

Cultivation

In plains, it is grown as a winter crop whereas on hills, as a summer crop from April onwards. Essential oil production and azulene content are influenzed by both temperature and light conditions. It prefers moist, moderately heavy, humus rich, neutral soils though its growth is satisfactory even in saline and alkaline soils having pH as high as 9.0.

Propagation is through seeds. Optimum temperature for seed germination is around 18-20°C. Seed rate is about 1 kg/ha. Seeds are sown in nursery bed and thinly covered with soil during September-October and usually germination occurs in 3-4 days and seedlings are transplanted when 6 weeks old, at a spacing of 30-40 cm depending on spreading habit of varieties grown.

Application of 20-25 tonnes/ha of well-rotten farm yard manure and 80:40:20 kg N, P_2O_5 and K_2O/ha are recommended for proper growth and yield. Normal soils require 3-4 irrigations whereas saline soils need frequent light irrigations during growing season. Generally, one or two weedings and hoeings are required for raising a good crop.

Flowering period is from February to April. Harvesting is done at full bloom stage at an interval of 10-15 days by plucking flowers. Generally yield is 4000-7000 kg fresh flowers from 4-5 harvests in a season. Presence of leaves, stem etc. reduce quality of oil obtained. Second and third harvests give higher yield. Flowers are dried under shade at 22-24°C by spreading in thin layers. Temperature controlled driers can also be used for this purpose. The dry yield varies from 1000 to 1500 kg/ha. Dried flowers are then packed and stored in moisture free environment.

Pest and disease management

Black bean aphids (*Aphis fabae*) are serious pests. An insect pest, *Nysius minor*, causes flower drop and another one, *Antographis chryson* causes defoliation. These pests can be effectively controlled by spraying suitable insecticides.

Extraction and utilisation

Essential oil is extracted by steam distillation of dry flowers for 4 hours at a pressure of 7 atmospheres/cm^2. Oil being very viscous, forms a deposit along inner walls of condenser. Hence flow of cooling water is to be frequently stopped till temperature rises sufficiently. Oil yield on dry weight basis varies from 0.3 to 1.3% depending upon location, soil fertility, climate, variety and growth conditions. A temperature range of 22-25°C favours maximum oil development in flowers. Average oil yield is 50-75 kg/ha. Oil content in German chamomile is 0.7 - 1.05%, Indian chamomile 0.5- 0.88% and Yugoslavian 0.34- 0.54%.

Physico-chemical properties of oil

Chamomile oil is viscous, with a characteristic aroma and bitter tonic flavour and intensely blue coloured immediately after extraction. On exposure to light, colour changes to green and finally to brown.

Specific gravity at 15°C	0.9326-0.9459
Acid value	18.7-31.7
Ester value	1.9-12.1
Ester value after acetylation	66.3-115.7
Solubility	Soluble in 90% alcohol

Essential oil contains, on an average 6% chamazulene, responsible for blue colour of oil and in addition, it contains azulene, farnesene, α-bisabolol oxide A and B and a dicyclo ether. Its healing properties are attributed to volatile constituent, α-bisabolol and flavonoid compound apinegin.

Cinnamon

Scientific name: *Cinnamomum verum* Presl; Syn: *C. zeylanicum* Bl.

Family: Lauraceae

Vernacular names: *Sanskrit*: Darusita; *Hindi*: Darucini; *Bengali*: Dalchini; *Malayalam*: Karuva; *Tamil*: Ilavargam; *Telugu*: Sapna lavanga; *Gujarathi*: Dalchini

Cinnamon, also known as true cinnamon or Ceylon cinnamon is an evergreen tree whose bark and leaves are strongly aromatic. It is one of the world's popular flavours and certainly the baker's most important spice. Bark exported as quills, is used as a spice or condiment, for flavouring cakes and sweets and in curry powders, incense, dentrifices and perfumes. Two types of essential oils are commercially extracted from cinnamon plants: bark oil from bark of tree and leaf oil from leaves and tender twigs. Bark oil is used in expensive perfumes, flavouring, confectionery, liquors and in pharmaceutical preparations, especially to mask unpleasant taste. Leaf oil is used in manufacture of cheaper types of perfumes used in soap, tooth paste and hair oil. In flavouring industry, it is used as a modifier. It is a cheap substitute for clove oil in seasoning. Eugenol, main constituent of leaf oil is used for synthesis of vanillin, major flavour component of vanilla beans. Essential oil possesses antimicrobial, fungitoxic, nematicidal and leech repelling activities.

Habitat and distribution

Cinnamon is indigenous to Sri Lanka, which is the largest producer and exporter of quill and bark oil of best quality. Plant reached Egypt and Europe by fifth century BC. It was introduced to Java in 1825 and has since been cultivated in India, Seychelles, Madagascar, Brazil, South East Asia and other tropical countries. In India, it is confined to lower elevations of Western Ghats in Kerala and lower Nilgiris of Tamil Nadu.

The plant

Cinnamon is classified in botanical division Magnoliophyta, class Magnoliopsida, order Magnoliales and family Lauraceae. The genus *Cinnamomum* has 250 species and many of them are aromatic and flavouring. Aromatic oils and oleoresins of commercial importance are extracted from a number of species of *Cinnamomum*. Most well-known among these are *C.verum, C. cassia, C. burmannii, C. loureirii* and *C. camphora*. *Cinnamomum verum* Presl.

is synonymous to *C. zeylanicum* Bl. with a chromosome number 2n = 24. Tree grows to a height of 7 to 10 m in its wild state and has deeply veined ovate leaves that are dark green on top and lighter green underneath. Being cross pollinated, wide variability is observed in population.

Cultivation

Several cultivars are known, but mostly sweet or honey types are cultivated extensively. For bark and bark oil, two varieties of cinnamon, namely Navashree (SL 63) and Nithyashree (IN 189) yielding 55.6 kg and 54.2 kg dry bark/ha/year, respectively are recommended for cultivation. These were selected based on their regeneration capacity, yield and quality. They have leaf oil recovery of 2.80% and 3.00% and eugenol content of 62% and 78%, respectively. A eugenol-rich leaf oil yielding cinnamon variety Sugandhini (ODC-130) with a leaf oil yield of 300 ml/tree/yr, oil recovery 1.6% on fresh weight basis (3.7% on dry weight basis) and eugenol 94% is recommended for leaf oil.

Cinnamon is a hardy plant which tolerates a wide range of climatic conditions. Wild trees are confined to tropical evergreen rain-forests upto 1800 m above MSL. Crop thrives well from 300-350 m and grows upto 1000 m above MSL. It flourishes in places with an annual rainfall of 1500-2500 mm and an average temperature of 27°C. A hot moist climate is highly suited for cinnamon cultivation. Proximity to sea, humid conditions and saltish water are good for crop. Sandy loam soil with admixture of humus or vegetative mould is the best for sweet and fragrant bark. Waterlogged and marshy areas are unsuitable.

Plant is propagated mainly by seed. True to type progeny is raised by cuttings of young 3-leaved shoots, division of old root stocks, air-layering of shoots and by tissue culture. Growth regulators are also used for inducing rooting and improving recovery. Application of a paste containing 1000 ppm IBA+2500 ppm NAA on girdled portion while layering is recommended.

Trees usually flower in January and fruits ripen 6 months later. Seeds are extracted from ripe fruits of selected trees with desirable characters like smooth bark, erect stem, easy peeling of bark, vigorous growth, freedom from pests and diseases and good qualities like sweetness, pungency and flavour. Seeds take 2-3 weeks time for germination. Seeds lose their viability fast and should be sown fresh after removal of pulp. About 94% germination is obtained by sowing seeds on third day after harvesting. At end of second week, germination is reduced to 52% and after 40 days there is complete loss of viability. Seeds are sown thickly in nurseries during June-July. When seedlings are four months old or reach a height of 15 cm they are transplanted into polythene bags of 30x15 cm size or into baskets. After a further period of 10-

12 months, when they have sufficiently hardened off, they are planted in main field.

Cinnamon being a cross pollinated crop, large variability is observed in progeny when propagated through seeds. This method is not suitable for propagation of elite types or varieties. True to type progeny is raised vegetatively by cuttings, air-layering or tissue culture. Root initiation and establishment in field are very difficult in cuttings. Tissue culture is also not feasible at farmers' level. Air-layering is very simple, effective and possible for every one. Optimum time for air-layering is April-May, when pre-monsoon showers are received. Six months to one year old twigs are suitable for layering. A ring of bark, 2 cm long, is removed from selected twig at the position where green colour of stem bark is turning from green to brown. A thick jute thread is tied around girdled portion to prevent patching up of bark from both cut ends. Girdled portion is then covered with moist sphagnum moss, coir pith or any suitable rooting medium held together by tying a plastic sheet around. A half cut is given at base of layers after one and a half months to induce stress and enhance rooting. Layers will be ready for separating from mother tree in three months when thick root growth is visible through transparent plastic cover. Since roots are geotropic in nature root growth is more on lower side of layer which is not exposed to direct sun light. Separated layers are carefully handled, plastic sheet is removed and planted in polythene bags, which are kept in shade for about a month and then transplanted.

Seedling or propagule is planted in main field at 2-3 m spacing. In India, under rainfed conditions, planting during June-July is ideal whereas for irrigated crop, planting during Oct- Nov. is recommended. Shading and irrigation are essential immediately after planting. During first year, seedling may be supplied with 20 kg cattle manure or compost and 20:20:25 g N, P_2O_5 and K_2O per annum which is gradually increased to 50 kg cattle manure and 200:180:200 g N, P_2O_5 and K_2O/tree/year for grownup plants of 6 years or more. This dose may be doubled for extra high yielding plants of age 15 years or more. Fertilizers may be applied in two equal splits during June-July and Oct-Nov. Application of Mussoriephos or Rajphos at 900 g/tree/year increases oil recovery by 75%. Foliar application of 4% urea increases leaf oil by 28%.

For first two years 3-4 weedings/year are required. Thereafter, two weedings in a year during June-July and Oct-Nov are sufficient. Mulching also reduces weed growth. Seedlings need irrigation till they are established, if there is long drought period.

Pest and disease management

Common diseases are leaf spot and die back caused by *Colletotrichum gloeosporioides*, grey blight by *Pestaloteopsis palmarum* and sooty mould caused by *Phragmocapinus betle* which can be controlled by spraying 1% Bordeaux Mixture or 0.3% mancozeb.

Extraction and utilisation

For preparation of quills, plants are harvested about 15 cm above ground level 3 years after planting when shoots have grown 2-2.5 cm in diameter and 1.5-2 m in length. Subsequent harvests can be done every 2-3 years. Generally, three germinating buds are allowed to grow as ratoon crop. Correct time for cutting shoots for peeling is determined by noting sap circulation between wood and corky layer. Peelers can judge this by making a test cut on stem with a sharp knife. If bark separates readily, cutting is taken immediately in early morning with sharp knife to prevent breaking and splitting of cut ends. The best time for peeling is when new flushes and leaves are hardened after a rainy season. Harvesting is normally done during October-November. Fully developed shoots harvested during rainy season gives good quality bark and high yield.

Leaves are used for extraction of leaf oil. Brown stem is scraped off and stems are cut into pieces of convenient length. Bark is split longitudinally and peeled off using peeling knife on same day of harvest or next day after softening. Peeled bark is rolled and made into cylindrical shape. Cylindrical pieces of bark dried in sun for 2-5 days and are packed in bundles. Dried cinnamon bark (quills) are graded on the basis of colour, taste, fragrance and weight and traded. First harvest yields 30-50 kg quills/ha/year. Better harvests are expected after 10 years when 180-200 kg of dried quills/ha/year is obtained.

About 75 kg/ha of quillings and featherings are also obtained additionally. These chips, featherings or trimmings of bark left after collection of quills are used for distillation and oil yield is 0.5-1.0% generally. Plants with intense purple flushes possessed higher bark oil content.

Cinnamon bark oil is light yellow in colour when freshly distilled. On storage, it becomes reddish. It has specific gravity (15.5°C) 0.950 to 1.030, refractive index (20°C) 1.565 to 1.599, optical rotation (20°C) 0° to 8° and its solubility in 70% alcohol is 10 volumes. Bark oil contains cinnamic aldehyde (60-75%), eugenol (10%), benzaldehyde, methyl amyl ketone, phellandrene, pinene, cymene, nonylaldehyde, linalool, cumin aldehyde, caryophyllene and esters of butyric acid.

Root bark oil (2-3% of root bark) is another variety of essential oil from cinnamon. It is colourless, lighter than leaf oil with specific gravity 0.994 and optical rotation +50° and contains most of terpenoids.

Cinnamon oleoresin is prepared by extracting cinnamon bark with organic solvents. Yield using ethanol is 10-12% and using benzene is 2.5-4.3%. Recently 1,1,2-trichloro-1,2,2-trifluroethane is also used.

Side shoots growing from base are cut to encourage growth of more side shoots till whole plant assumes shape of a bush so as to maximise leaf production. For extraction of leaf oil, leaves and tender twigs are harvested every year during May and/or November. Wilting of harvested leaves in shade for 24 hours reduces bulk and increases oil recovery. Steam distillation for 4-6 hours gives an oil yield of 0.5 to 0.8% generally on a commercial scale.

Leaf oil is heavier than water, yellow to yellowish brown in colour with a slight camphoraceous odour resembling that of clove oil due to presence of 70-95% of eugenol. Leaf oil has specific gravity (15.5°C) 1.065, refractive index (20°C) 1.530 to 1.545, optical rotation (20°C) -1° to +3°, acid value 14.0-15.7, ester value 4.7-16.7 and its solubility in 70% alcohol is 10 volumes. Leaf oil contains approximately α-pinene 0.2%, 1, 8-cineole 1.65%, *p*-cymene 0.35%, α-ylangene 0.25%, linalool 1.5%, caryophyllene 1.85%, α-humulene 0.2%, α-terpineol 0.15%, piperitone 0.1%, safrole 0.65%, cinnamaldehyde 1.3%, cinnamyl acetate 0.8%, eugenol 87%, acetoeugenol 1.0%, cinnamyl alcohol 0.60%, benzyl benzoate 2.68% and traces of over 15 other compounds.

Citronella (Java)

Scientific name: *Cymbopogon winterianus* Jowitt
Family: Poaceae (Graminae)
Vernacular names: *English*: Java citronella grass; *Hindi*: Ganjri; *Bengali*: Khaddi, Banchi; *Malayalam*: Kamakshi pullu; *Marathi*: Ilsadhana; *Punjabi*: Khavai; *Tamil*: Kamachi pullu; *Telugu*: Kamakshikasavu

Citronella is a tufted perennial grass, leaves of which on distillation give an yellowish-brown essential oil with citrus odour. Two types of citronella are identified, viz. Java and Ceylon citronella, former gives superior oil of commercial importance while latter yields inferior quality oil. Java oil serves as a starting material for extraction of geraniol and citronellal which can be converted into aroma chemicals such as citronellol, hydroxy citronellol, synthetic menthol and esters of geraniol. These find extensive use in soap, perfumery, cosmetic and flavouring industries throughout world. Soaps, soap flakes, detergents, household cleansers, insecticides and other technical products are often perfumed exclusively with this oil. Citronellol is used in many perfumery blends of soap and cosmetic industries when rosaceous notes are required. Hydroxy citronellol is a key ingredient in compounding and in floralizing perfume materials. Citronellol esters like formate and acetate are used in a wide range of fragrances. Oil is also used in manufacture of deodorants, mosquito repellent creams and allied products. Oil of Ceylon citronella is employed chiefly for perfuming low priced technical preparations such as detergents, sprays, polishes and insect repellents. Spent grass can be used as a source of raw material for cellulose pulp and paper production by using sulphate, sulphite and cold caustic soda.

Habitat and distribution

Citronella thrives well under tropical and sub-tropical conditions. It requires abundant moisture and sunshine for good growth. Rainfall of 100-150 cm influences growth of plant, yield and quality of oil favourably. However, its distribution is more important than total rainfall. In areas where rainfall is less, plant can be grown with supplemental irrigations. Though, the plant grows under a wide range of soil conditions, sandy loam uplands with abundant organic matter is the most suitable. Heavy clay soils and sandy soils do not support good growth. Crop is very sensitive to water-logging. It grows under a wide range of pH (5-8) though 6-7 is ideal.

Citronella grass has now been widely distributed throughout tropics comprising Sri Lanka, Java, Central America, Guatemala, Taiwan, Brazil, East Africa, Congo, Malagasy Republic, Seychelles, India and West Indies. In India, it is cultivated in States of Assam, West Bengal, Uttar Pradesh, Maharashtra, Karnataka, Tamil Nadu, Gujarat, Arunachal Pradesh, Manipur, Mizoram, Meghalaya, Nagaland and Tripura. Production of Java citronella exceeds other essential oils. World production of citronella oil is about 7000 tonnes/year; major producers being Taiwan, China, Indonesia and Guatemala. Chief importing countries are USA, UK, West Germany, Japan and Hong Kong. At present, world production of citronella oil is approximately 5000 tonnes out of which India produces about 300-350 tonnes of oil.

The plant

Citronella is a tufted aromatic perennial herb with fibrous roots. Culms stout, erect, over 0.60 to 1.8 m, terete, smooth and shining, leafy, glabrous at nodes; leaf blades linear, gradually tapering to a long membraneous, acuminate tip, up to 1 m long, 1.5 cm wide and yellow-green above; inflorescence, a very large decompound panicle, spreading loose, over 30 cm long, erect and finally drooping.

There are two types of citronella.

C. winterianus Jowitt (Java citronella)

C. nardus Rendle (Ceylon citronella)

Both these types are originated from Mana grass of Ceylon. Java citronella, which is called '*Maha pengeri*' in Ceylon, is result of selection from Ceylon citronella.

C. winterianus: Oil from this type is superior. Plants are over 2 m in height. Clumps are stout, erect, terete, leafy and glabrous at nodes. Leaf blades linear, with long acuminate tip and serrate margins. Ligules are scarious. Inflorescence is a very large compound panicle. Racemes 20 mm long, one sessile and other pedicelled with two lower spikelets, homogeneous, male or neuter, remaining pairs in both racemes heterogeneous. Ovary superior with one carpel, unilocular. Stigma bifid, lateral and leathery. Ovule single and erect. Java citronella flowers profusely in South India and at higher altitudes in hills of North-Eastern India. In plains of North and North-Eastern India, it flowers only sporadically. Viable seeds are not formed because of irregularities in meiosis and therefore, the species can be propagated only vegetatively.

C. nardus: This type is cultivated in Sri Lanka. It is hardier and can be grown on poor soils. Leaves are smaller than that of Java type and yield an inferior oil with a lower citronellal content. Plant is a robust, stoloniferous perennial grass, 0.5-1.0 m high, leaves broad, panicles large and mostly compound, spikelets awnless, lanceolate and flattened on back.

Cultivation

Due to pressure on land for production of major food and agricultural crops, following areas possess potential for commercial cultivation of citronella.

a. Large sized holdings where farmers are not able to manage their land due to management cost in intensive and diversified agriculture.

b. Where production of traditional crops is affected by wild/pet animals.

c. In areas where frequent watch and ward is difficult.

d. In agroforestry systems (up to 30% shade).

Java citronella is more extensively and commercially cultivated than Ceylon type, though former is less hardy compared to latter. Grass is propagated only vegetatively by slips which are obtained by dividing well grown clumps. Clumps are separated in a manner that each slip contains 1-3 tillers. Roots and leaves are trimmed off before planting. About 50 slips are obtained from an year old healthy clump. Slips should be obtained from at least 6 months old plantation. Slips from plantations of above 4 year are not desirable. 'Jorlab C2', 'RRL JOR-3-1970', 'CIMAP/Bio-13' and 'CIMAP/73-1' are improved high oil yielding varieties available for cultivation. With onset of monsoon, land is brought to fine tilth by ploughing and harrowing. Ridges and furrows or beds are made. Slips obtained from healthy vigorously growing plants are planted during June-July at 60-90 cm spacing and 10 cm deep. Care should be taken to avoid waterlogging in field. Delay in planting results in drying of slips and poor establishment and plant population. FYM is applied (10 tonnes/ha) before planting. A fertilizer dose of 200 kg N, 80 kg P_2O_5 and 40-80 kg K_2O/ha/annum is recommended for optimum growth and yield. Better results are obtained by applying N in 4 equal split doses at an interval of about 3 months. P and K are applied fully as basal. Irrigation is required within 24 hours of planting, if there is no rain. Supplemental irrigation is not required in areas with 2000-2500 mm rainfall, well distributed over the year. However, in drier months irrigation may be provided for better yield. Depending on weather and soil conditions, 8-10 irrigations are required in dry areas during rain free period. Field is to be kept weed free till a complete cover of crop is obtained. Weeding in inter row spaces can economically be achieved using cultivators between rows. Earthing up is done after about 4 months of planting and again after every harvest as citronella rootstock has a tendency to work out of soil by itself.

Crop is ready for first harvest after about 9 months of planting and subsequently at an interval of 3 months. Harvesting is done by sickle above first node at 20-45 cm above ground. Generally leaf blades are cut and sheath are left out. Flowering should be discouraged as it causes aging and reduces life span of plantation. Harvesting can be done 4 times a year. Generally, Java crop once planted yields profitable returns for 4-5 years and needs replanting only afterwards. Ceylon citronella remains productive for 10-15 years. Uprooting of plantation after its life span and rotation with any legume species is recommended. Horse gram, cowpea and sunhemp are good rotational crops.

Pest and disease management

Pests

Adults of termite species, *Microtermus obesi* (Isoptera: Termitidae) damage roots of plants through out year and plants get dried. Grubs of white grub, *Holotrichia consanguinea* (Coleoptera: Melolonthidae) feed on roots. Dirty white larvae of shoot borer, *Chilo infuscatellus* (Lepidoptera: Pyralidae) with fine longitudinal stripes bore into stems of young plants causing dead heart. Chlorpyriphos 20EC @ 600-800 ml/ha with irrigation controls termites, white grubs and shoot borer. Yellowish-green aphids, *Macrosiphum miscanthii* (Homoptera: Aphididae) feed on leaves in summer months. Aphids can be controlled by application of any contact insecticide. Army worm, *Mythimna separata* (Lepidoptera: Noctuidae) and grass hopper, *Colemania sphenariodes* (Orthoptera: Acrididae) are defoliators usually seen in citronella. Spraying of malathion 50EC (0.05%) or quinalphos 25EC (0.05%) can effectively control pests, if damage is severe.

Diseases

Leaf blight: This disease caused by two fungal pathogens, *Curvularia* and *Colletotrichum* is one of the most serious and devastive diseases, which appears after onset of monsoon in all the Java citronella growing areas in India. In Assam, the disease is caused by *Curvularia eragrostidis*. Initial symptoms of disease appear as small dark reddish brown necrotic spots all over infected leaves which enlarge in size and coalesce to form big elongated necrotic lesions that result into severe blight symptoms. Blight caused by *C. andropogonis* in severe form results in great economic loses to crop. Fungus produces profuse spore masses in the form of brown dot like structures in centre of old necrotic lesion that serve as a source of perpetuation of disease. Incidence of disease is controlled by prophylactic spraying of mancozeb (0.3%) or zineb (0.3%) at an interval of 15 days. Leaf blight caused by *Colletotrichum graminicola* affects the crop occasionally in Karnataka. This disease can also be controlled by application of above fungicides.

Lethal yellowing: It is a devastating diseases of Java citronella. Incidence of disease occurs during July depending on onset of monsoon. It appears in epidemic form during August and September. Distinctly visible yellow leaves with poor growth due to fungal infection on roots are major symptom. In advanced stages, infection caused by pathogen spreads from roots to basal portion of stem resulting into premature death and drying of large populations of plants in irregular patches. Disease is caused by *Pythium aphanidermatum*. Reproductive organs such as oogonia, antheridia and oospores, characteristic of *P. aphanidermatum* are abundantly seen in roots of infected plants. Incidence of disease is very high in low areas and poorly drained soil because abundant moisture favours infection. In initial stages, pathogen attacks succulent root and causes severe damage to root system. As a result, nutrient uptake is hampered leading to lethal yellowing. Disease could be easily managed by slip treatment and three foliar sprays either with 0.1 % solution of Ridomil-Mancozeb. Fungicides, carbendazim and topsin–M are also effective in reducing mortality but had very little effect on lethal yellowing caused by *P. aphanibermatum*.

Collar rot and wilt: Java citronella is generally infected by this disease during March-April. It affects 20-30% of plants in severely infected fields. Initially, infected plants produce crinkled, curled and yellow leaves. In advanced stage of infection, a symptom of collar rot leading to premature drying and death of infected plants occur. The pathogen, *Fusarium moniliformae* is identified as cause of disease. Infected plants show white cottony growth of pathogen under moist conditions. Foliar sprays either with carbendazim or topsin–M @ 0.2% is highly effective for control of this disease.

Extraction and utilisation

Harvested grass is wilted in shade for a short time and steam distilled within 24 hours. Oil yield varies with season, soil fertility and distillation efficiency. On an average, 0.8-1.2% of oil is recovered from grass and oil yield is about 100 kg/ha during first year and 150 kg/ha during subsequent years. Yields of 200-250 kg/ha/yr can be obtained under favourable conditions with good management.

At present, world production of citronella oil is approximately 5000 tonnes, bulk of it is produced in Taiwan, Guatemala, Malaysia, Brazil, Ceylon, India, Argentina, Equador, Madagascar, Mexico and West Indies. India produces about 300-350 tonnes of oil. Assam, Arunachal Pradesh, Karnataka, Maharashtra, Tamil Nadu, West Bengal and Uttar Pradesh are leading states.

Physico-chemical properties of oil

Specific gravity at 15°C	0.887-0.895
Optical rotation	-0° 35' to -5°6'
Refractive index	1.4685 - 1.4728
Total geraniol	82.3 - 89.4%
Aldehydes and cintronellal	28.8 - 43.9%
Solubility in 80% alcohol	1-2 vol.

Java citronella oil contained citronellal 32-45%, geraniol 12-18%, citronellol 11-15%, geranyl acetate 3.8%, citronellyl acetate 2-4%, limonene 2-5%, elemol and other sesquiterpene alcohols 2-5%, α-elemene and α-cadinene 2-5%, and traces of cubebene, calamenene, bourbonene, bisabolene, eugenol, methyl eugenol, isopulegol, nerol, linalool, geraniol, methyl heptanone, myrcene and α-pinene. Good quality citronella oils contain more than 38% of citronellal, 16% of geraniol and 12% of citronellol.

Davana

Scientific name: *Artemisia pallens* Wall.ex DC.
Family: Asteraceae (Compositae)
Vernacular names: *Sanskrit*: Davanam; *Hindi*: Davana; *Tamil*: Davanam; *Kannada*: Davana

Flowering top of davana yields an essential oil which is extensively used in high grade fine perfumes. Oil is used for flavouring cakes, pastries, tobacco, beverages, sausages and preserved products. Laves form an important component of garlands and bouquets.

Habitat and distribution

Davana is a native of South India. Though, it is not systematically cultivated in its home country, she holds key position in production of Davana oil. Annual production is about 2 tonnes. It grows common in Kashmir valley, Simla, Nainital hills, Karnataka, Tamil Nadu, Uttar Pradesh and Andhra Pradesh. In the production of davana essential oil, India holds key position and production is about 2 tonnes/annum. Essential oil of davana acquired considerable reputation in international trade. From Karnataka, Tamil Nadu and Andhra Pradesh oil is exported to USA and Japan.

The plant

It is a delicate, erect and branched annual herb 45-60 cm tall and covered with greyish white tomentum. Leaves are alternate, exstipulate, petiolate, lobed; inflorescence capitulum, axillary, peduncled to sessile, heterogeneous having bisexual disc florets in centre and pistillate ray florets on periphery which are yellow glabrous. Involucre, 2 or more, seriate, ovate to elliptic-linear; inner florets 5-lobed, bisexual; stamens 5 with free epipetalous filaments; style bifid.

Cultivation

Davana being a delicate plant cannot withstand heavy rains. It prefers light drizzles, bright sunshine and a mild winter with no frost and heavy morning dew during growing season. Cloudy weather and rains during flowering and seed ripening stages adversely affect yield. Crop grown during November gives maximum herb and oil yield. However, crop can be grown round the year for use in garlands and bouquets. Plant grows on various types of soils ranging from sandy loam to medium black soils, but humus rich red loam soils are ideal.

Plant is propagated by seeds. Seeds are short-viable and hence cannot be stored for long. Transplanting is generally practised in the crop. A nursery area of 500 m^2 sown with about 1.5 kg seeds is sufficient for planting one hectare. Seeds are mixed with fine sand, broadcast over nursery bed, covered with a thin layer of sand and watered regularly. Seeds germinate in about 3-4 days. When seedlings are 10-12 cm tall, they are transplanted to main field at 15x7.5 cm spacing. Before transplanting, 12-15 tonnes of well decomposed FYM and 40 kg/ha each of phosphorus and potash are incorporated into soil. N is applied at 120 kg/ha in 4 equal splits, 3 for main crop and 1 for ratoon crop at 15 days interval. After transplantation of seedlingts plots should be irrigated lightly till seedlings are well established. Two weedings are carried out in main crop and one in ratoon crop. Application of GA_3 (200 ppm) increases herbage and essential oil yield per unit area. Crop is harvested during February-March (about 120-125 days after sowing) when a large number of flower buds start opening.To obtain maximum essential oil yield, plants should be harvested when about 50% of them have come to flowering stage. Harvesting is done by cutting whole plant with sickle at a height of 10 cm from ground. Yield of fresh herbage and flower heads is around 12-13 tonnes/ ha.

Pest and disease management

Crop is often subject to damping off at tender early stage, particularly in nursery. This disease, caused by *Rhizoctonia* species is common during cloudy and rainy period. Hence, adjusting of sowing time is important for avoiding disease.

Extraction and utilisation

Harvested herb is dried in shade for 2-3 days. Dried herb is steam distilled for a period of 6-8 hours for extracting essential oil. Flower heads contribute the major portion of oil. Percentage of oil content on two days air-dried basis shows that flower heads contain 0.3-0.4% of oil and in general, an oil recovery of 0.2% is achieved from whole plant. Oil yield is 12-15 kg/ha.

The best yield and quality of davana essential oil are obtained from plants grown in summer.Physico chemical properties of oil are: Specific gravity at 20°C 0.9605, refractive index at 20°C 1.4880, optical rotation 35, acid value 2-4, ester value 52.9 and solubility in 70% alcohol 10 vol.

Davana oil contains davanone, fenchyl alcohol, cinnamyl cinnamate, caryophyllene, cadinene, linalool, dehydro-α-linalool, davanafuran, isodavanone, dihydrorosefuran, n-alkanes, hydroxy davanone, geraniol and nerol. Davanafurans are responsible for characteristic odour of davana oil though they constitute only 0.8% of oil.

Eucalyptus

Scientific name: *Eucalyptus* spp.
Family: Myrtaceae
Vernacular names: *English*: Blue gum tree; *Sanskrit*: Nilanirgasa; *Hindi*: Yukaliptas; *Malayalam*: Yukkali; *Tamil*: Karupuramaram; *Kannada*: Taila

Eucalyptus is an essential oil yielding tree which has perfumery, industrial and medicinal uses. Essential oil is used in soap and cosmetic industries. It is an effective substitute for Java citronella oil and a source of citronellal for manufacture of citronellol, hydroxy citronellal and menthol. Oil is used in germicides and disinfectants to improve odour. It is used as an antiseptic especially in treatment of infections of upper respiratory tract and in certain types of skin diseases. It is used as a stimulating expectorant in chronic bronchitis and as a vermifuge against hookworm. It is internally administered or inhaled with steam for asthma and respiratory disorders. Besides oil, it yields high volume of wood in a short duration of 6 to 8 years having a variety of uses. Wood is used as mine props, railway sleepers, for paper manufacture and house construction.

Habitat and distribution

Eucalyptus is an exotic genus, which has shown its adaptability to a wide range of environments and is extensively grown almost throughout India. It is indigenous to Australia. It is widely distributed over tropical and subtropical countries in Asia, Africa and America. In India, it is grown in Kerala, Tamil Nadu, Karnataka, Meghalaya and Uttar Pradesh. One of the most important characteristics of Eucalyptus is plasticity, which enables trees to grow well in difficult and different climatic conditions. Regular cultivation of *E. globulus* started in India from 1856 onwards. About 80% of country's total production of *E. globulus* oil is produced in Palani and Nilgiri Hills.

Eucalyptus plants prefer tropical and subtropical climate but grow satisfactorily in temperate regions as well. They are frost sensitive in early stages. *E. citriodora* grows in plains upto 600 m while *E. globulus* prefers an elevation of 1300-2700 m. Former grows well on poor gravelly soils in rainfed areas but can be grown on any soil, whereas latter is chiefly grown on moisture rich loamy soils.

The plant

There are about 700 species of genus Eucalyptus and all species, except four, are endemic to Australia. Some of them possess medicinal value or volatile oils having fragrances varying from camphor, thymol, peppermint to rose and lemongrass. *Eucalyptus citriodora* Hook. and *E. globulus* Labill. are the most common species. *E. citriodora* commonly known as citron (lemon) scented gum or spotted gum is a tall, graceful tree, 25-40 m high, with a crown of leaves and branches at top. Bark smooth, white to pinkish. Distinct cotyledonary, seedling, juvenile, intermediate and adult leaves are observed at various stages of plant. Adult leaves alternate, lanceolate; inflorescence usually axillary, corymbose; panicle umbels 3-5 flowered on terete 5-7 mm long peduncles; flowers pedicellate; calyx tube hemispherical or cylindrical; anthers adnate; fruits ovoid, truncate, contracted at edge, rim thin and valves opening downwards.

Eucalyptus grows to a height of 55 m. Bark blue-grey; inflorescence axillary, usually solitary, but occasionally in 3-flowered umbels on a very short or rudimentary peduncle; flowers sessile, operculum, flattened, hemispherical; fruit sessile, globular to broadly conical, 4-ribbed and warty.

Cultivation

Plant is propagated by seeds only. Seeds can be sown directly in field or seedlings can be raised in nursery during February-September. About 100 g seeds are sufficient for raising seedlings for one hectare. Seeds are soaked for 24 hours before sowing to enhance germination. Soaked seeds are mixed with sand and then sown in lines in raised beds. Seeds germinate in 7-8 days and seedlings are ready for transplanting in polybags in about 30 days. Fully grown seedlings get ready for planting in main field in about five months time. They are transplanted to the mainfield at 70-100 cm spacing. 10-12 tonnes/ha of organic manure and 120:60:60 kg/ha of N, P_2O_5 and K_2O may be applied. Weeding is required during initial period until plants are established well.

Harvesting is done when weather is clear and oil content in leaves is maximum. Accordingly, harvesting leaves during February, April, July and October is recommended. In Kerala, harvesting twice a year, in May and November, is practised. Plants can be pollarded to promote vigorous sprouting of side shoots. Fresh shoots sprout in about four weeks after pollarding, which are ready for harvesting after 4-5 months. First pruning is done at 30-45 cm above ground and subsequent ones at 75-90 cm above ground.

Pest and disease management

Pests

Important pests of *E. citriodora* are termites, leaf cutting ants, Eucalyptus snout beetles, weevils, scale insects, cicadas, thrips and wood borers. Termites damage roots and bark of stem. Drenching soil with chlorpyriphos EC with irrigation water is the easy management measure against termites. Lindane 5% dusting can also be resorted to. Soil should not be kept dry for long periods. Other defoliators can be controlled by application of any contact fungicides like mercaptothion (0.2%), if required.

Diseases

Blight caused by *Cylindrocladium scoparium* affects seedlings, leaves and stems of *E. citriodora* and at times the disease becomes severe. Drying and defoliation of leaves are also caused by *Physalospora latitans*. Leaf spot in younger leaves is caused by *Petalotiopsis funera* in *E. globulus*. Fungus *Fusarium orthoceros* attacks seedlings of *E. citriodora* at hypocotyl level. *Colletotrichum* infection on leaves is also noticed. These diseases can be controlled by spraying/drenching of carbendazim (0.1%), mancozeb (0.3%) or 1% Bordeaux Mixture. Other fungal diseases reported in eucalyptus are stem-end rot (*Pseudophaeolus baudonii, Diaporthe spp., Corticium salmonicolor*) and basal stem canker (*Phytophthora nicotinae*) which normally do not become serious. Three types of viral diseases viz. tobacco mosaic virus (TMV), little leaf virus and leaf crinkle virus are noticed in *E. citriodora*. Destruction of affected plants is the only remedy for reducing spread of disease.

Extraction and utilisation

Oil is obtained by steam distillation of either fresh or dried leaves. Leaves are collected from pruned branches of four to five year old trees. After pruning, trees are allowed to rejuvenate for 18 to 24 months and again same process is repeated. Abscised and fallen dried leaves also can be collected and distilled. Leaves are steam distilled for 2-3 hours soon after harvesting to avoid loss of oil through evaporation as well as deterioration of its quality during storage. Though oil recoveries as high as 5% is reported, 1-2% yield is frequently achieved in *E. citriodora*. The highest oil yield is obtained from top leaves. This oil has better solubility in alcohol and higher cineole content than oil obtained from lower leaves. A 6-8 year old tree yields 30-60 kg of leaves/year which gives 0.5-1 kg of oil. About 100 kg oil/ha is obtained from a four year old plantation. A well maintained plantation produces about 400-500 kg oil/ha from seventh year onwards. On an average, about 100 tonnes of timber is also obtained from 8-10 year old plantation. Essential oil of eucalyptus obtained by steam distillation of mature leaves from different species differs considerably in composition. Based on their chemical composition, oil can be broadly divided into three groups; cineole rich, citronellal rich and phyllandrene rich oils. *E. globulus, E. australiana* and *E.smithini* produce cineole rich oil; *E. citriodora* gives citronellal rich oil and *E. dives* yields an essential oil rich in phyllandrene. While cineole rich oil is used for medicinal purposes, citronellal rich oil has industrial uses. Essential oil of *E. globulus* is important as there is no cheap alternate source of cineole. Synthetic routes to citronellal, hydroxyl citronellal and phyllandrene have reduced importance of other oils and thus cultivation of species which produce these oils.

Physico-chemical properties of oil

E. citriodora oil is colourless to light yellow with a grassy verbena odour.

Specific gravity at 15°C	0.8635-0.8765
Optical rotation	-3° to +3°
Refractive index at 20°C	1.4511-1.4570
Aldehydes expressed as citronellal	65-85%
Ester value	12-60
Ester value after acetylation	230-292
Solubility in 70% alcohol	1.3-1.5 vol.

Approximate composition of oil is α-pinene 0.1-1.9%, β-pinene 0.4-1.9%, myrcene 0.1-0.6%, limonene 0.4-7.1%, 1,8 cineole 1.1-17.9%, *p*-cymene 0.3-0.9%, γ-terpinene 0.2-0.9%, terpinolene 0.1-0.8%, citronellal 26.7-82.6%, linalool 0.3-0.9%, iso-pulegol 4.7-29.8%, citronellol 5.1-13.4%, citronellyl acetate 0.41-0.7% and caryophyllene 0.3-3.9%.

E. globulus oil is colourless to light yellow with camphoraceous odour and with following properties.

Specific gravity at 20°C	0.9065-0.9155
Optical rotation	-9° 39' to +5° 27'
Refractive index at 20° C	1.463-1.466
Acid value	0.18-1.04
Saponification value	8.90-12.0
Saponification value after acetylation	17.00-21.68

Major constituents of oil are cineole (60-70%) caryophyllene, camphene, sabinene, myrcene, *p*-menthane, β and γ -terpinene, fenchone, and β-thujone, citral, verbenone, iso-amyl alcohol, trans-pino carveol, borneol, myrtenol, eudesmol, thymol, bornyl acetate, caproic acid, piperitone and globulol.

Geranium

Scientific name: *Pelargonium graveolens* L. Her. *ex* Ait; Syn: *P. roseum* Willd., *P. intermedium, Geranium radula*
Family: Geraniaceae

Pelargoniums and geraniums are generally known as geraniums. Geranium oil blends well with all kinds of scents, floral and oriental bouquets and is extensively used in perfumery and cosmetic industries. It is widely used for scenting soaps due to its stability in alkaline medium. Oil is also used for production of Rhodinol used in manufacture of perfume compounds.

Habitat and distribution

It is believed that geranium was first introduced to Kenya either from South Africa by Cape Dutch settlers or from India by engineers engaged to build Uganda railway. Geranium oil is produced in a number of east and central African countries, principally Kenya, but oil production has virtually ceased except for a small quantity for a domestic consumption in South Africa. Total world production of geranium oil is about 300 tonnes per year and major producers are China, Egypt, Morocco, and Reunion Island of South Africa. Crop was introduced in high altitude areas of Southern India about four decades ago. In India, geraniumu is widely cultivated and oil produced in plains of Bangalore, Hyderabad and hills of Uttaranchal and also cultivated in north Indian plains as an annual winter-summer season crop.

The plant

Geranium is a small, much branched perennial aromatic herb of about 1.3 m height, naturally forming a dense spreading bush. Stem is cylindrical, pubescent and woody at base. Leaves are fragrant, mint scented and opposite, 6.5–13.5 cm, villous, highly divided with 5-7 revolute. Petioles are long sturdy and subtended by stipules. Inflorescence is axillaries with small umbels of 3-7 pink, rosy or pale purple flowers. Fruit is long and pointed, but normally without seeds. Some cultivars produce seeds. Seed is small, oblong-ovoid and brownish; testa is hard with attached long hairs. Viability is usually very poor and germination erratic.

Cultivation

A Mediterranean mild climate with low humidity warm winter, mild summer temperature and an annual rainfall of 1000-1500 mm are ideal for crop. It grows successfully at an altitude of 1000-2100 m. Well drained porous soils are suitable for its cultivation. Saline, alkaline and damp soils are unsuitable. Plant is propagated by stem and root cuttings. Terminal cutting roots earlier than middle and basal cuttings. IAA is better than IBA for inducing rooting. Rooted cuttings are raised in nursery during November-January and transplanted after two months at 60x40 cm spacing in main field after applying well decomposed FYM or compost. Irrigating field on previous day of transplanting is compulsory to get better yield. Apply FYM at 10-12 tonnes/ ha. Inorganic fertilizers like phosphate and potash are applied at 40-60 kg/ ha as basal while N is applied up to 200 kg/ha/year in six equal splits to cover 3 harvests. Application of micronutrients such as Cu at 20 kg/ha/year and Mo at 3 kg/ha/year in 3-4 split doses is beneficial. Irrigation is provided daily for first 3-4 days on alternate days till two weeks and weekly thereafter. Crop requires weeding at 20 and 40 days after planting. Crop is ready for harvest after 4 months from transplanting when leaves begin to turn light green and exhibit a change from lemon like odour to that of rose. Green leafy shoots are harvested with a sharp sickle. Use of sharp sickle minimizes damage to crop while harvesting. Crop after harvesting is maintained by hoeing, fertilizer application and irrigation as per schedule. It puts forth fresh shoots, grows fast and reaches harvesting stage in about four months. Three harvests can be taken per annum and crop remains in field for 4-6 years. Yield of fresh herbage/ha/year is about 15 tonnes. 'Bourbon' and 'CIMPAWAN' are high yielding genotypes. Quality of oil obtained from these varieties holds good value in international market. Variety 'CIMPAWAN' yields 20-25% more herb and good quality oil.

Pest and disease management

Geranium is affected by several fungal diseases and insect pests causing severe herbage and oil yield losses.

Pests

White ants (Termites): Different species of termites such as *Microtermes obesi, Odontotermes obesus, Termes taprobanses* (Isoptera: Termitidae) damage roots and stems of plants. Adult termites which are brownish dirty white in colour feed on plant parts throughout year. Application of chlorpyriphos 20EC at 600–800 ml/ha with irrigation water keeps termites under control.

Root knot nematodes: Attack of root-knot nematodes *Melodogyne incognita* and *M. hapla* are common in this crop. These are controlled by applying carbofuran 3G @ 1.0 kg a.i/ha.

Diseases

Leaf blight: This fungal disease is caused by *Colletotrichum gleosporioides* and *Alternaria alternata.* Dark brown minute necrotic lesions appearing on leaf margin expand towards midrib resulting in complete necrosis and blighted appearance of leaves. During rainy season 60-70% damage may occur. Disease can be controlled by three to four rounds of foliar application of captafol @ 0.2% or chlorothalonil @ 0.3% or mancozeb @ 0.1% at fortnightly intervals, if damage are severe. Soil drenching with carbendazim 1 g/l or hexaconazole 1 g/l may be done. Benomyl 0.2% has to be sprayed before and after harvesting and one week after harvesting of crop. Use only sharp implements for harvesting and avoid unnecessary disturbance to soil and excess irrigation.

Root rot and wilt: Rhizoctonia solani, Botryodiplodia theobromae and *Fusarium oxysporum redolens* cause drooping of lower leaves initially and plant collapses within a week. Uprooted plants show severe rotting of roots. Basal portion of stem also rots and turns black at later stages. Planting in already infected soils and water logged areas should be avoided to get a disease free crop. Chlorothalonil or mancozeb drenching at 0.3% and pretreatment of rooted cuttings with 0.3% mancozeb are control measures for this fungal disease.

Rhizoctonia solani also causes rot in stem cuttings by which fresh leaves droop and basal portion of cuttings becomes black and plant dies off in a week. Dipping cuttings in 0.3% mancozeb or chlorothalonil and drenching nursery beds are control measures.

Extraction and utilisation

Harvested herb is immediately taken up for distillation. Steam distillation gives better quality oil as compared to hydro distillation. Distillation takes 3-4 hours. Quality and yield of oil are better if the crops are harvested at right maturity stage. On the basis of origin there are three plant types of geranium cultivated in India such as Algerian, Bourbon and Egyptian. Bourbon or reunion type yields the best quality essential oil with high market value. Reunion geranium oil possesses a very strong, heavy rose like odour, occasionally slightly harsh and minty. Volatile oil is present mostly in leaf blades and there is practically no oil in woody stem. In large scale distillation, oil recovery varies from 0.1 to 0.15% on fresh weight basis and average oil yield is 18 to 20 kg/ha/year. Maximum oil yield of 60 kg/ha was recorded. The oil should be filtered to remove the impurities that might have come into the oil during distillation. Then the oil can store in thoroughly clean, dry self sealing aluminium containers. To avoid deterioration of the quality of the oil the oil should be fiiled up to the brim level of the containers. Geranium oil possesses strong, somewhat rose like odour which is reported to improve with age if properly stored. The major chemical composition of the essential oil is reported to be iso-menthome 5.2 to 7.2%, linalool, 3.96 to 2.90%, guaia 6.9%, diene 0.15 to 4.4%, citronellyl formate 1.9 to 7.55%, citronellol 19.28-40.23% and geraniol 6.45-18.4%.

The physico-chemical properties of the reunion geranium oil has specific gravity at 15°C 0.888-0.896, optical rotation -7° 40′ - -13°50′, refractive index at 20°C 1.461-1.468, acid number 1.5–12, ester number 50–78, estercontent calculated as geranyl tiglate 21-33%, ester number after acetylation 206 to 233, total alcohol content, calculated as geraniol 67 to 77.6% and solubility usually clearly soluble in 2 to 3 volume of alcohol, after separation of paraffin crystals on addition of more alcohol.

Lavender

Scientific name: *Lavendula angustifolia* Linn.
Family: Labiatae (Lamiaceae)

Lavendula is a small genus which includes 28 species of perennial aromatic herbs, sub shrubs and shrubs. Out of these, only three species *L. angustifolia* (True lavender), *L. latifolia* (Spike lavender) and *L. hybrida* (Lavendin) are extensively used for extracting essential oils.

True lavender is one of the most important perfumery materials. Spike lavender which has a mixed fragrance of lavender and rosemary is inferior in quality to lavendin and is used in several lower grade perfumes. Lavendin is a hybrid of *L. angustifolia* and *L. intermedia* and its oil combines characteristics of those of its parents. Though it does not stand quality of true lavender, oil of lavendin is the most traded material because hybrid species is easier to cultivate and gives high yield of oil.

Lavender is widely used in perfumery. It possesses a delightful refreshing odour of its own and blends well with many other perfumes to produce a variety of premier blends. Lavender water, a hydro-alcoholic solution of lavender oil is one of the most popular toilet articles in Europe. It is also used for scenting toilet soaps, talcum powders and disinfectant formulations.

Habitat and distribution

True lavender is distributed in mountain districts of southern Europe bordering western half of Mediterranean and extending across eastern coast of Spain, France, Switzerland, North Italy and North Africa. Major producers of lavender oil are France, Italy, Russia, Hungary, UK, Bulgaria, Australia, China and USA. Spike lavender is distributed in Europe, Spain, Mediterranean region, Balearic isle, France, Italy and Balkan Peninsula. It is cultivated in Dalmatia, France, Italy and Spain.

The plant

Three species of *Lavendula* are generally identified. *L. angustifolia* Miller ssp. *angustifoila* syn. *L. officinalis* Chaix, *L. vera* DC is a shrub, 50-80 cm in height, tomentose. Leaves usually dimorphic, basal and axillary ones small, dense, grey tomentose; primary leaves of young shoots larger up to 6 cm, margin much revolute. Peduncle unbranched and erect. Spike compact or interrupted especially at base. Bracts are membraneous or hyaline, ovate or very broadly ovate to obovate, acuminate or apiculate, almost glabrous, pubescent or hispid especially on median and lateral nerves which are conspicuous. Bractioles very small, linear, brown or absent. Calyx up to 5 mm long, marginal teeth short or rounded or margin entire, posterior appendage oblate or rotund, often deep purple. Corolla tube is nearly double the calyx, length up to 1 cm, lobes large.

Lavendula latifolia Mill. Syn. *L. spica* DC is a shrub of up to 60-80 cm, leaves dimorphic linear, lanceolate, narrowly elliptic to spathulate, much attenuated at base; primary leaves of young shoot up to 6 cm long, becoming glabrescent, margins scarcely revolute, basal and axillary leaves rather small and fastigiate. Peduncles typically branching, spreading and often very long. Spikes often interrupted, compact, rather slender. Bracts linear to lanceolate, acute, tomentose or hispid, equal to calyx or slightly longer. Bract on median nerve alone is conspicuous. Bractioles linear, green or grayish. Calyx up to 5 mm long, marginal teeth obtuse or rounded, posterior appendage elliptic or ovate. Corolla tube about 7.5 mm long, lobes small.

Lavendula intermedia Emeric ex Loiset syn. *L. hybrida* Revr is a shrub 40-50 cm, leaves dimorphic, 4-5 cm long, linear lanceolate and narrowly spathulate, attenuated at base, margin may or may not revolute, basal leaves smaller and fastigiate and densely gray. Peduncles less branching, spikes mostly at anterior end and compact. Bracts membraneous, linear lanceolate and prominently veined. Bractioles ovate to obovate, scariosus, green or grayish. Calyx up to 5 mm long and marginal teeth obtuse or rounded, posterior appendages oblate or rotund. Corolla tube 7 mm long, lobes big.

Cultivation

Lavender is a temperate and photophilous plant. It can be grown successfully on arable lands at very high altitudes. It does well only in those areas which have cold winter and cool summers. Crop requires a light, well aerated, dry, fertile and calcareous soil. As plant is highly sensitive to high moisture and hence water logged soils are unsuitable. However it withstands drought and frost very well. Lavender grows very well on sloppy lands; it has a very deep root system as well. Hence crop is useful for checking erosion to a great extent.

Lavender can be propagated vegetatively using cuttings. Although seed propagation is cheap and quick, commercial plantations are established using cuttings. Cuttings are obtained from annual and biennial branches of mother plantation during October-November in plains and in February-March on hills. Nursery beds of 20-25 cm height are made. A 5 cm layer of 1:1 mixture of organic manure and sand is spread over it and finally covered with a 3-4 cm layer of clean river sand. Cuttings of 8-10 cm length are planted on beds at a spacing of 4 to 5 cm between them. Rooted cuttings are transplanted in main field at a spacing of 1.20-1.40 m between rows and 35-40 cm within row. First harvest is taken three years after planting. Flowers are harvested when crop is in full bloom. Inflorescence at a length not greater than 12 cm is collected during day time when weather is warm and sunny. Productivity of plantation declines after 3-4 years when it is time for replanting.

Extraction and utilisation

Oil of lavender is extracted by steam distillation. Period of 90 min. is required fresh flowers. For production of high quality lavender oil, stainless steel distillation vessels, which are operated by steam from a boiler are required. Essential oil content of different varieties varies from 0.5-1.1% with 0.8% as average. A yield of 50-80 kg of essential oil per hectare can be obtained.

Lavender oil with about 50% lavendulyl esters is the best quality of oil and is used in preparation of high grade perfumes. Second grade oil (38-42% esters) is used for preparation of lavender waters, toilet waters and Eau-de-cologne and low grade (30-35% esters) in scenting of soaps and talcum powders.

Physico-chemical properties of oil

Lavender oil is a pale yellow to amber coloured mobile liquid. It has a characteristic lavender odour with a slight cineolic or camphoraceous note. Lavender oil from different sources vary in physico-chemical composition as well as odour characteristics and this variation is attributed to differences in species, latitude, longitude and altitude of area where it is cultivated, age of crop, time of harvesting and distillation and mode of distillation. Physico-chemical characteristics of a typical lavender oil are given below.

Specific gravity (20°C)	0.891-0.899
Optical rotation (20°C)	-7° to -3.5°
Acid value	Maximum 1.0
Ester value	100-137
Solubility (20°C)	One volume soluble in 4 volumes 10% ethanol

Lavander oil contains linalool 29.5%, linalyl acetate 33%, camphor 7%, 1,8-cineole 5.5%, terpeni-4-ol 3.25%, borneol 2.25%, lavendulyl acetate 2.25%, limonene 1%, *cis*-β-ocimene 1%, *trans*- β-ocimene 0.5% and lavendulol 0.5% on an average.

Lemongrass

Scientific name: *Cymbopogon flexuosus* (Nees ex Steud) Wats.
Family: Poaceae
Vernacular names: *Sanskrit*: Bhustarah; *Hindi*: Gandhatran; *Bengali*: Gandhabena; *Malayalam*: Injippullu; *Tamil*: Vasanapullu; *Kannada*: Majjigehallu; *Telugu*: Nimmagaddi; *Gujarathi*: Lilacha; *Punjabi*: Khavi

Lemongrass is a tropical perennial plant which yields aromatic oil. The name lemongrass is derived from typical lemon-like odour of essential oil present in shoot. Lemongrass oil of commerce is popularly known as Cochin oil in world trade, since 90% of it is shipped from Cochin port. The state of Kerala in India had monopoly in production and export of lemongrass oil. Annual world production of lemongrass oil is around 1000 tonnes from an area of 16000 ha. In India, it is cultivated in about 4000 ha and annual production is around 250 tonnes. Crop is extensively cultivated in poor, marginal and waste lands and also along bunds as live mulch. Well-ramified root system of plant helps in soil and water conservation.

Lemongrass leaves are widely used as a lemon flavour ingredient in herbal teas and other formulations. Lemongrass is commonly used in Asian cooking. When Thai food was embraced in US, lemongrass became a household name. A little experimentation with this delightfully fragrant herb is all, it takes to realize that it can be used in many more ways than just in Asian dishes. A simple syrup made by steeping lemongrass in a mix of equal parts of hot water and sugar can be used to enhance taste of fruit salads or to make home made soda by mixing it with seltzer. A blend of lemongrass, garlic, ginger and oil will be stable in freezer during winter. This paste can be fried until fragrant and then cooked down with a can of coconut milk (strain to remove tough lemongrass fibres) for delicious sauce for noodle, vegetable or seafood dishes.

Lemongrass oil is used in culinary flavouring. It is used in most of the major categories of food including alcoholic and non-alcoholic beverages, frozen dairy desserts, candy baked foods, gelatins and puddings, meat and meat products and fat and oils. It is used to improve flavour of some fish and can be used to flavour wines and sauces.

Habitat and distribution

Lemongrass is distributed in Africa, Indian sub-continent, South America, Australia, Europe and North America. In India, they grow wild in all regions extending from sea level to an altitude of 4200 m. Several species are endemic to India. East Indian Lemongrass grows wild in India and is cultivated well in Kerala, Assam, Maharashtra and Uttar Pradesh. It is also distributed in Guatemala and China. West Indian lemongrass originated either in Malaysia or in Sri Lanka. It is widely distributed throughout tropics and is grown in West Indies, Guatemala, Brazil, Congo, Tanzania, India, Thailand, Bangladesh, Madagaskar and China. Jammu lemongrass is confined mostly to North Indian states such as Jammu and Kashmir, Sikkim, Assam, West Bengal and Madhya Pradesh. Lemongrass is cultivated in large scale at Chinnar wildlife sanctuary in Western Ghats of India.

The plant

Lemongrass belongs to family *Graminae* (Poaceae) and genus *Cymbopogon*. Generally, three species are identified.

C. flexuosus (Nees ex Steud) Wats. (2n=20, 40)

It is known as East Indian, Cochin or Malabar grass. *C. flexuosus* is a tufted robust perennial grass of about 2 m height. Leaves are linear and lanceolate. It flowers freely. Inflorescence is very large and a highly branched terminal drooping panicle bearing paired spikes on tertiary branches. Spikes bear spikelets in pairs of which one is sessile and other pedicellate. The sessile spikelet is an awned bisexual floret where as the pedicellate is an awnless staminate floret. Under this species, two varieties or types are identified based on colour of stem.

C. flexuosus var. *flexuosus* is the red grass. Stem and leaf sheath are reddish or purple. It is recognized as true lemongrass and is commercially cultivated. The essential oil contains more than 75-80% citral, exhibits good solubility in alcohol and hence is superior in quality.

C. flexuosus var. *albescens* is the white grass characterized by white colour of stem. Plant is normally seen wild. Essential oil contains less than 65-70% citral, exhibits poor alcohol solubility and hence considered inferior in quality.

C. citratus (DC) Stapf. (2n=40, 60)

It is known as West Indian or American lemongrass. It is a stemless perennial grass with numerous stiff tillers arising from short rhizomatous rootstock, making large tussocks. It seldom flowers under cultivation. Leaf blade is narrow, linear, glaucous, drooping with scabrous margin, ligule truncate, inflorescence rarely produced, a large loose panicle; spathe bracts long and narrow, sessile spikelets, awnless, linear and lanceolate. Essential oil contains 74-76% citral and exhibits poor alcohol solubility.

C. pendulus (Nees ex Steud) Wats.

It is Jammu lemongrass and is white stemmed and dwarf in nature. Plant is frost resistant and suited to Sub-Himalayan areas of North India. Essential oil contains around 75-80% citral and exhibits medium solubility in alcohol.

Cultivation

Cymbopogon flexuosus and *C. citratus* flourish in sunny, warm, humid conditions of tropics. In Kerala, lemongrass grows well between 900 and 1250 m from mean sea level. Both species produce the highest oil yield/tonne of herbage where the rainfall averages 2500-3000 mm annually. *C. citratus* is more drought tolerant. In areas where rainfall is poor, it can be grown with supplemental irrigation. Day temperature of 25-30°C is optimum for maximum oil production, with no extremely low night temperature. Short periods above 30°C have little general effect on plants, but severely reduce oil content.

Lemongrass flourishes in a wide variety of soil ranging from rich loam to poor laterite. In sandy loam and red soils, it requires good manuring. Calcareous and water-logged soils are unsuitable for its cultivation. Both species can be grown on a range of soils and it appears that good drainage is the most important factor. Plants growing in sandy soils have higher leaf oil yield and citral content. Although *C. flexuosus* flourishes in well drained sandy loams, in India, it is grown in almost all types of land available from very light sandy soil to upland laterites. Soils of pH 5.5 to 7.5 are utilized. *C. citratus* is more commonly grown on soils with higher acidity than *C. flexuosus*. In India, the highest herb and oil yields/hectare of *C. flexuosus* are obtained in soils of pH 7.5. Lemongrass grows and produces average herbage and oil yields on highly saline soils. In pot trials, *C. flexuosus* grown in soils with electrical conductivity of 11.5,10 and 5.5 mmhos/cm showed no significant reduction in herb and oil yield and citral content was unaffected by increasing salinity levels up to 15 mmhos/cm.

Lemongrass varieties released for cultivation are Sugandhi, Pragati, Praman, RRL-16, CKP-25, RRL-39, Kavery, Krishna, SD-68, GRL-1 and SB-9.

Sugandhi (OD-19): Released from Aromatic and Medicinal Plants Research Station (AMPRS), Odakkali, Kerala, India. A red stemmed variety adapted to a wide range of soil and climatic conditions and the most popular in India. Plant grows to a height of 1-1.75 m with profuse tillering, yielding 35-40 tonnes/ha/year herb containing 0.3% oil (125 kg/ha) with 80-85% citral under rain-fed condition.

Pragati (LS-48): Evolved through clonal selection from OD-19 at Central Institute of Medicinal and Aromatic Plants (CIMAP), Lucknow, India. It is tall growing with dark purple leaf sheath, adapted to North Indian Plains and 'Tarai' belt of subtropical and tropical climate. Average oil content is 0.63% with 86% citral.

Praman (Clone 29): It is evolved through clonal selection at CIMAP, Lucknow and belongs to species *C. pendulus*. It is a tetraploid type with profuse tillering. Leaves are erect and medium in size. Oil yield is 227 kg/ha/annum with 82% citral content.

RRL-16: It is evolved from *C. pendulus* and released for cultivation from Regional Research Laboratory (RRL), Jammu, India. Average yield of herb is 15 to 20 tonnes/ha/annum giving 100 to 110 kg oil. Oil content varies from 0.6 to 0.8% and citral content is 80%.

SD-68: Developed by S.C. Datta, using ionizing radiation yields up to 375 kg of oil/ha/year with a citral content of 90-92 %.

RRL-39: Released from RRL, Jammu.

Kavery and Krishna: Released from CIMAP Regional Station, Bangalore, India.

Chirharit: A high yielding variety, developed by systematic breeding for genetic improvement at Pantnagar, Chirharit, India. It is frost resistant and the essential oil contains 81% citral.

Lemongrass germplasm consisting of about 406 accessions is maintained at AMPRS, Odakkali. There are 17 other types in germplasm in which major constituent of oil is not citral.

Lemongrass is generally propagated through seeds. Seed is mixed with dry river sand in the ratio of 1:3 and sown in field at the rate of 20 to 25 kg/ha. Alternatively, seedlings can be raised in a nursery in one-tenth of area of main field and transplanted after 45 days. This method which requires 3-4 kg seeds/ha of main field is ideal for uniform stand and better growth of plants. Small plantation of lemongrass can be established by planting of slips.

Cympogon flexuosus is propagated through seeds while *C. citratus* is propagated through division of clumps. Propagation through vegetative means from selected clones is better as seed propagation tends to cause considerable genetic heterogeneity resulting in deterioration of yield and oil quality and clonal proliferation plays a very important role in propagation of lemongrass.

Lemongrass seeds have dormancy of a few weeks and they lose viability in a few months. Seeds collected during January-February are usually sown in nursery during April-May. Germination is very poor if sown after October. For one hectare of land, 1000 m^2 nursery has to be raised. Area is made to fine tilth by repeated ploughing. Beds of 1 to1.5 m width and convenient length are prepared. Seeds are uniformly broadcasted on beds at 3-4 kg/ha and covered with a thin layer of soil. Seed bed is irrigated frequently. Seeds germinate in 5-7 days.

Seedlings raised in nursery beds are transplanted in field at 6-7 leaf stage. 50-70 days old seedlings are planted during monsoon. A spacing of 30 cm x 30 cm with a plant density of 111 000/ha is recommended. A wider spacing of 60 cm x 45 cm for seedlings and 90 cm x 60 cm for slips are recommended for fertile, irrigated land under North Indian conditions.

Spent lemongrass compost at 10 t /ha and wood ash at 2 t /ha, which are obtained as by-products of grass distillation are applied at time of bed formation. Lemongrass requires 275 kg N, 50 kg P_2O_5 and 175 kg K_2O/ha/ annum. Under rainfed conditions of Kerala, application of 100 kg N in 3 to 4 split doses was optimum though a response up to 200 kg was recorded. Application of 50 kg/ha each of P_2O_5 and K_2O as a basal dose gave encouraging results in West Bengal. It is recommended to apply 60:45:35 kg / ha N, P_2O_5 and K_2O basally and 60 kg N in 3 to 4 splits/annum as top dressing during growing season as an optimum dose. It also responds well to application of copper, iron, calcium and sulphur. It was reported from CIMAP, Lucknow that a lower dose of boron (2.5 ppm) in combination with chloride salts (chloride salinity) is beneficial for crop.

In chromate overburdened soil, application of lime at 6 tonnes/ha and fertilizer at 100 kg N, 50 kg P_2O_5 and 50 kg K_2O/ha produced higher plant height, tiller number and herb yield in *C. pendulus*.

Soluble nitrogen fraction and total carbohydrate content increased essential oil content. Pattern of formation of citral in *C. flexuosus* oil revealed that constituents increased up to reproductive phase and then declined, it again increased after post-reproductive phase of plant. Optimum application of fertilizers increase citral content of oil. Excess fertilizer application is undesirable as it promotes more vegetative growth and oil with less citral content.

In case of drought, crop should be irrigated every alternate day for about a month after planting. It is recommended that 4 to 6 irrigations are given during February to June under North Indian conditions for optimum yield. Soil moisture regimes maintained at 0.80 IW: CPE ratio significantly increased crop growth, herbage and essential oil yields. Quality of essential oil is not affected by soil moisture regimes.

The first 25-30 days after planting (or harvest) is crop-weed competition period. For a good establishment of crop, field should be kept weed free for initial period of 3-4 months after planting. Once the crop is well established, it can compete with weeds.

Generally, 2-3 weedings are necessary in a year. Among herbicides, diuron at 1.5 kg ai/ha and oxyfluorfen at 1.5 kg ai/ha are effective for weed control. A significant control of dicot weeds was observed with application of 2-4-D (sodium salt). Spraying paraquat at 2-2.5 l/ha in 500 l of water immediately after cutting the grass is an excellent method of weed control.

Under rainfed conditions, the field gives a dried appearance during December–May. The dry grass and stubbles of crop is set on fire in May, prior to onset of monsoon. This practice kills termites attacking crop stubbles and also helps to rejuvenate old clumps.

Plant does not tolerate shade and oil yield is drastically reduced when crop is grown under diffused light. Studies at AMPRS, Odakkali indicated poor tillering, lean and lanky growth and reduced oil yield when crop is grown as intercrop in coconut gardens; oil content was also found reduced by 20%. In contrast, intercropping in cinnamon plantation which is regularly pruned for extraction of bark and leaf oil was profitable. In new plantations of cashew, mango and coconut, lemongrass is cultivated during initial 4 to 5 years of plantation establishment. *C. citratus* is seldom intercropped or under-planted. An interesting method of integrating *C. flexuosus* into plantations of other crops was proposed for Bangladesh, but not widely implemented. *C. citratus* is under-planted in young rubber plantations in Malaysia and elsewhere to help defray cost of plantation establishment. Lemongrass was suggested for crop diversification in semi-arid regions.

Harvesting is done by cutting grass 10 cm above ground level with sickles. Number of harvests in a year depends on climatological factors like temperature, rainfall and humidity and level of soil fertility. Generally crop thrives the best in humid condition. Cutting can begin as soon as nights dews have evaporated from plants, as wet grass left for later distillation quickly ferments. Sunny days are preferable, since cloudy and misty conditions tend to depress leaf oil content. First harvest can be taken at 90 days after planting and subsequent harvest at 50-55 days interval up to 5-6 years from same crop. During first year of planting, three cuttings are obtained and subsequently 5-6 cuttings per year. Harvesting season begins in May and continues till end of January. An herbage yield of 10-15 tonnes/ha/harvest may be obtained.

Lemongrass kept for seed purpose is not cut as yield of seeds from plants subjected to regular harvest is very low. Generally, plant flowers during November-December in plains and mature seeds are collected during January-February. A healthy plant produces 10 to 20 g of seeds. The whole inflorescence is cut and dried in sun and seeds are collected by thrashing against floor or beating with sticks. Fresh seeds are recommended for use in raising a plantation since seeds lose viability beyond six months of storage. Seed germination is very poor till May, increases up to July and thereafter decreases. Germination is meager beyond October.

Pest and disease management

Pests

A few pests are reported in this crop. Infestation by spindle bug (*Clovia bipunctata*) has been observed in Kerala and severe damage by a stem boring caterpillar of *Chilotrea sp.* under North Indian conditions is reported. Spraying mercaptothion (0.2%) controls insects. Nematodes like *Tylenchorhynchus vulgaris, Rotylenchulus reniformis, Helicotylenchus spp.* and *Pratylenchus spp.* also infect the grass.

Diseases

Common diseases and their causal agents are given below. These leaf diseases can be managed by prophylactic sprays of zineb @ 3g/l thrice, at intervals of 15 days or application of 0.2% copper oxychloride or 1% Bordeaux Mixture.

Disease	Causal organism
Little leaf (malformation of inflorescence)	*Balensia sclerotica* (Pat) Hohnel
Leaf spot (eye spot)	*Helminthosporium sacchari, H. leucostylum, Drechslera victoria* and *D. helm*
Leaf spot	*Curvularia andropogonia* (CLS)
Leaf spot	*C. veruciformis, C. trifolii* and *Collitotrichum graminicola*
Leaf spot and clump rot	*Fusarium equiseti* and *F. verticillium*
Leaf blight	*Curvularia andropogonia* (CLB)
Leaf blight	*Rhyzoctonia solani.*
Grey blight	*Pestalotiopsis magniferae*
Smut	*Tolyposporium christensenni* and *Ustilago andropogonis*
Root rot	*Botrydiplodia theobromae*

Helminthosporium cymbopogi cause very serious disease in low lands of Guatemala. Brown top disease causes browning and curling of affected leaves. This is a physiological disease resulting from low water content of grass at end of dry season. Symptoms of rust disease of lemongrass causing elongated, stripe like, dark brown lesions on both sides of leaf surfaces have been described. Causal organism is *Puccinia nakanishikii*. Root segments of lemongrass were heavily infested with multiple vesicular arbuscular mycorrhiza (VAM). Moreover, brown septate hyphae of non-mycorrhizal fungus also co-existed with VAM in 50% of root segments. Burning of stubbles in summer is practised in some areas to ward off pests, diseases and weeds.

Leaves become smaller in size and flowers in inflorescence get converted to very small leaves in case of little leaf caused by virus. Removal and destruction of such plants should be resorted to, to reduce spread of disease and avoid collecting seeds from such plants.

Extraction and utilisation

Lemongrass oil is collected by steam distillation of herbage. Distillate on cooling separates out into a layer of oil, floating over bulk of water. For obtaining good quality oil, steam distillation in stainless steel units is preferred at a steam pressure of 18-32 kg/cm^2 in boiler. Grass is distilled either fresh or after wilting. Wilting herbage prior to distilling reduces moisture content and increases oil recovery. Drying in sun reduces oil recovery but has little effect on oil composition. Generally, Clevenger apparatus is used for distilling small quantities (up to 1.0 kg) of herb in laboratory. Large field scale distillation units are fabricated to distill 500 kg or more of the herb at a time. On an average, herbage of *C. flexuosus* contains 0.2-0.4% oil and oil yield is 100-125 kg/ha/year. Distillation being a high temperature process, yields an oil with burnt note. Also it is devoid of volatile fractions. An oil of softer note is yielded by solvent extraction. However, the process is more expensive than steam distillation. Residue obtained after extraction of oil is called spent grass. It can be used as cattle feed fresh or after ensilaging. It can be used for mulching or manuring crops as such or after composting. In some plantations in India, spent lemongrass after drying is used as a fuel for distillation. It is also a cheap packing material. Spent grass on an average contains N 0.74%, P 0.07%, K 2.12%, Ca 0.36%, Mg 0.15%, S 0.19%, Fe 126.73 ppm, Mn 155.82 ppm, Zn 35.51 ppm and Cu 56.64 ppm.

Oil of lemongrass is a viscous liquid, yellow to dark yellow or dark amber in colour turning red on prolonged storage. Presence of water imparts a turbid appearance. Whole oil is mainly used as a source of citral. Differentiation of lemongrass oils into West Indian and East Indian in trade has no geographical significance, as oils from both species are produced in these areas. However, the West Indian oil has less citral and more myrcene than East Indian oil. Although both oils have a pronounced fresh lemony fragrance, odour of East Indian is stronger. East Indian is considered fresher, lighter and sweeter.

Morphological characters like plant height, number of tillers/plant and number of leaves/plant is significantly correlated with essential oil yield/ plant. Maximum elimicin content as a major chemical constituent of oil is at flowering stage. Among the physiological characters, a significant correlation was observed between essential oil content and crop growth rate (r=0.6018) as well as net assimilation rate (r=0.9474).

Oil of lemongrass is chemically reactive. The terpene mixture undergoes a series of complex reactions when exposed to air and sunlight. It is slowly converted into a dark coloured viscous resinous substance on keeping. However, if stored in aluminium or stainless steel vessel with out contact of air, water and light, quality of oil is stable for long periods.

East Indian lemongrass oil contains 75-85% of aldehydes consisting largely of citral. Other constituents in oil are linalool (1.34%), geraniol (5.00%), citronellol, nerol (2.20%), 1,8 cineole, citronellal (0.37%), linalyl acetate, geranyl acetate (1.95%), α-pinene (0.24%), limonene (2.42%), caryophyllene, β-pinene, β-thujene, myrcene (0.46%), β-ocimene (0.06%), terpenolene (0.05%), methyl heptanone (1.50%) and α-terpineol (0.24%).

Essential oil of *C. citratus* contains approximately α-pinene (0.13%), β-pinene, δ-3-catrene (0.16%), myrcene (12.75%), dipentene (0.23%), β-phellandrene (0.07%), β-cymene (0.2%), methyl heptanene (2.62%), citronellal (0.73%), β-elemene (1.33%), β-caryophyllene (0.18%), citronellyl acetate (0.96%), geranyl acetate (3.00%), citral b (0.18%), citral a (41.82%), geraniol (1.85%), elemol (1.2%) and β-caryophyllene oxide (0.61%).

Average composition of *C. pendulus* oil is pinene (0.19%), camphene (0.01%), β-pinene (0.16%), car-3-ene (0.04%), myrcene (0.04%), dipentene (0.35%), phellandrene (0.3%), p-cymene (0.36%), methyl heptanone (1.05%), citronellal (0.49%), linalool (3.07%), β-elemene (0.7%), β-caryophyllene (2.15%), citronellyl acetate (0.72%), geraniol acetate (3.58%), citral b (32.27%), citral a (43.29%), geraniol (2.6%), elemol (2.29%) and β-caryophyllene oxide (1.56%).

Two isomers of citral constitute bulk of lemongrass oil. Citral is separated from oil by fractional distillation and used as a starting material for synthesis of a number of industrially important products. Citral has a citrus flavour. Geraniol, linalool and citronellol are the most important acyclic terpene alcohols that can be separated from lemongrass oil and used as flavour and fragrance substances. In flavour compositions, geraniol is used in small quantities to accentuate citrus notes. Nerol is used for bouquetting citrus flavours. Citronellol too is added for bouquetting purposes to citrus compositions. Pinene is an important starting material in fragrance and flavour industry.

A total extract of lemongrass comprising of volatile and non-volatile components imparting flavour and aroma to the product can be prepared by subjecting herb to extraction with a suitable solvent or a mixture of solvents. Oleoresin that results will be a concentrated wholesome product with better storage characteristics.

Leaves of lemongrass can be used as a source of cellulose in manufacture of paper and cardboard. Reduction in root-knot nematode disease was observed in soil amended with leaves of *C. flexuosus*. In the Caribbean, lemongrass is primarily regarded as a fever reducing herb (especially where there is significant catarrh). It is applied externally as a poultice to ease pain and arthritis. In India, a paste of leaves is smeared on patches of ringworm.

Citral is the starting material for preparation of ionones. α-ionone is used in flavours, cosmetics and perfumes. β-ionone is used for synthesis of vitamin A. Citral b, the most common constituent of oil, could be a good inhibitor of β-glucuronidase. The oil has other uses as bactericide, as insect repellant and in medicine. Antimicrobial cream, Wisprec made of *Ocimum sanctum* and *C. citratus* remains intact in its activity up to three years from date of manufacturing. Its mosquito repellent activity lasts 2-3 hrs. It exhibits significant antifeedant and larvicidal activity against *H. armigera*. Oil of *C. citratus* caused egg hatch inhibition. It is effective against storage pests. Whole oil has fungicidal properties to plant and human pathogen and is potentially anticarcinogenic. Essential oils from *C. citratus* were tested for their cytotoxic activity against P_{388} leukemia cells. It also exhibited antioxidant activities comparable with α-tocopherol and butylated hydroxyl toluene. It retards mould growth in butter cakes thereby increasing storage life. Oil of *C. pendulus* is used for preparation of antibacterial drug trimethoxyprim. Z-asarone, a component of oil is used as antiallergic compound. It is used for development of designer beverages and blends of oils with desired odour characteristics. It strengthens stomach, stimulates appetite, promotes digestion, and regulates nervous system and vascular expansion. It is a stimulant, antiseptic, febrifuge, carminative, diuretic, anti-inflammatory, anti-diabetic and useful against rickets.

Linaloe

Scientific name: *Bursera delpechiana* Poisson ex Engl.
Family: Burseraceae

B. delpechiana is one of important essential oil yielding plants introduced to India from Mexico in 1920. Natives of Mexico call this tree linaloe or Indian lavender tree and its wood berries and leaves yield essential oil called linaloe oil. It is used as raw material for extraction of linalool. Because of its stability to alkali, oil is a highly valued perfumery material and is used in numerous perfumes, cosmetics and in scenting soaps and transparent soaps in particular. It is also used in flavoring food and beverages. Linaloe oil is extensively used as a fixative for high grade perfumery and cosmetic products.Berry oil resembles *Bois de Rose* and can be used as a fixative in perfuming lily, lavender, cananga and soaps.

Habitat and distribution

In India, Bursera is grown in Karnataka, Maharashtra and Andhra Pradesh. Area under crop in India is around 800 ha, out of which 450 ha is government owned. Linaloe plant was introduced first near Bangalore in Karnataka. Thatguni estate in Karnataka which is 19 Km away from Bangalore, is perhaps the only place in the whole world where linaloe trees were cultivated as a regular crop. Most of oil was sold in Indian market and rest was exported to London. Trees grow wild in Mexico and oil is mainly obtained from wood of more than 20 years old. Oil obtained in India is mainly from matured berries and is called as linaloe berry oil.

The plant

Bursera is a deciduous dioecious tree. Medium sized male tree has a longer trunk and grows to about 7.5 m. Female tree attains about 6 m height and has larger leaves measuring up to 18cm in length. Leaves imparipinnate, serrate or nearly entire; flowers green, pentamerous; calyx small, lobes rounded; petals valvate in bud; stamens 10, nearly equal; ovary hairy surrounded by a broad crenate disk; fruits fleshy and dark green berry turning to reddish brown as they mature and fall off.

Genus *Bursera* has 45 species. Important species are as follows.

1. *B. gummifora*: American gum tree or Indian birch or birds's tree. An infusion of its leaves is occasionally used as a substitute for tea.
2. *B. icilarita* : Tree bears edible aromatic fruits
3. *B. javanica*: Juice of the fruits and leaves are used.
4. *B. leptopholes*: Mature fruits are edible.
5. *B. serratea*: Pulp is edible and also yields essential oils.
6. *B. semarubia*: Commonly called Rumb or "West Indian birch"; its leaves used as a substitute for tea.
7. *B. penicilliata*: Every plant part contains essential oils.
8. *B. kluggis*: It yields triterpenes, which have anticancer properties.

Cultivation

Species *B. delpechiana* comes up well even in stony, gravelly and very shallow soils, where other species are not likely to grow well. Linaloe is a hardy tree which grows in high tropical climates where maximum temperature varies between 36°C and 38°C. It grows well in areas with rainfall of 500-900 mm. Plant prefers frost free condition. Plant favours dry summer months where average blossom rains during April-May increase incidence of flowering and fruiting. It grows up to an altitude of 760 m. In well drained sandy soils with neutral pH, plant attains good height with spreading branches. Water logging causes cracking of stems and finally wilting away to death.

Plant is propagated by stem cuttings and seeds. Stem cuttings are usually used, as seeds do not germinate easily and germination is very poor. A good and healthy mother plant which is of more than five years age is preferred to obtain cuttings. Best season for raising cuttings is during February-March. Cuttings of 0.5-1 m length are usually taken. While preparing cuttings for planting, side branches up to a height of about 25 cm from the base are removed. Shoot cuttings are planted in earthen pots of about 12 cm in diameter and 25 cm height containing good garden soil with adequate drainage. Pots containing cuttings are kept under shade and watering is done at appropriate intervals to keep the soil moist. First sprout appears in about 30-40 days and healthy root system develops in about 4-6 months. About 90% recovery is obtained. Rooted cuttings are transplanted to main field during rainy season in pits of 60 cm^3 at a spacing of 6 m. Irrigate for a week and then fill with top soil and compost and press well. Watering is done at regular intervals. Since Bursera is dioecious, for every eight female plants one male plant has to be planted to meet pollination requirements. Plants establish in field very quickly. They start shedding leaves during November and are completely without foliage till late March when new flushes appear. They are very hardy and once established do not need much care. No serious pests and diseases are noticed. Linaloe plant raised from cuttings set fruits first year itself while those from seeds take about 5 years for fruit set. New flush of leaves along with flower buds appear in April. Berries start setting by May and mature by July-August when they are harvested, dried and dehusked. One kg of dried husk is obtained from 5-6 kg of fully mature berries. Propagation through seeds by birds and natural regeneration underneath plant is also noticed. Other means of natural regeneration is by coppicing. It coppices well, but it is not encouraged because of spreading nature of coppices. It rarely produces suckers. Plant gives a successful crop by third or fourth year. Earlier fruit setting is not allowed to induce adequate vegetative development. Picked or fallen fruits on clean ground are collected by sweeping floor of plantation. Bursera is a cross pollinated crop and pollination in this crop is mainly aided by insects. Maximum fruit set of 26% was recorded under hand pollination while it was 16% under open pollination.

Extraction and utilisation

Almost all parts of linaloe tree contain aroma. Mexicans distill wood while Indians use outer husks of berries. Husk oil yield is much less; 1.8% as compared to 2.5-3.0% obtained from wood in Mexico. Oil is extracted by steam distillation. Care should be taken to see that steam pressure in boiler is between 1-1.5 kg. Berries can be steam distilled either fresh or dry. Period for distillation of fresh fruits is five hours and dry ones are 20-25 hours for dry. Actual duration varies with equipment, quantity and steam pressure. Period of distillation is appreciably less when fresh fruits are distilled. Still should not be filled up to brim as husks swell during distillation. Air dried husks yield 10-14% oil.

Yield of oil per hectare depends on locality, age of trees, number of trees, proportion of male trees, cultural operations and rain during fleshing season.

One hectare of plantation containing 300 plants (10 years old) yields about 1500 kg of berries. Yield of husk is 18% of berry weight, accounting for 255 kg/ha. Oil content of husk is 10%, thus yielding about 25 kg/ha of oil. Fresh fruits yield s1.5-2.5% oil. Wood oil is distilled from wood of 40-60 year's old trees which yield 7-12% oil while younger trees yield 2.5-3% oil. Seed oil produced in India is known as *Mysore Linaloe oil or Indian Lavender oil*. Leaf oil yield is 0.15-0.25%.

Physico-chemical properties of oil

Husk oil is a light coloured mobile liquid much lighter than water and soluble in 70% ethanol. Approximate composition of oil is methyl heptanol

1.5%, linalool 47.7%, linalyl acetate 40.8%, sesquiterpene and other viscous substances 8%. Mexican oil contains 60-75% linalool. Leaf oil has a sweet wafting odour and it contains 65-70% linalyl acetate. Reported physico-chemical properties of oil are as follows.

Specific gravity at 20°C	0.8952
Refractive index at 20°C	1.4658
Optical rotation at 20°C	1.5°
Acid value	0
Saponification value	130
Saponification value after acetylation	274

Indian linaloe oil contains methyl heptenol 1.5%, linalool 47.7%, linalyl acetate 40.8%, sesquiterpene and other viscous substances 8%. Other constituents reported are myrcene, limonene, trans-linalool oxide, cis-linalool oxide, neryl acetate, cis-and trans-2,6,6-trimethyl-2-vinyl-5 acetoxytetra hydropyrane as additional constituents.

Mints

Scientific name: *Mentha* sp.
Family: Lamiaceae
Vernacular names: *Sanskrit*: Pudina; *Hindi*: Pudina; *Bengali*: Podina; *Malayalam*: Pudina; *Tamil*: Pudina; *Kannada*: Chetamarugu; *Telugu*: Pudina; *Gujarathi*: Pudina; *Assamese*: Theihadun

Mints are aromatic perennial herbs with quadrangular stem and bearing leaves with essential oil present in glands located in subcuticular region. Among various types of mints, only Japanese mint is cultivated in tropics or subtropics with a cooler climate. Anti microbial properties of menthol mint essential oil enhances shelf life of edible products and grains. It is generally cultivated as a primary source of menthol and is extensively used in betel (pan) related industries. Mints are also used for flavoring toothpastes, candies, beverages, confectionery, chewing gums, pan parag, and mouth washes and for scenting shaving creams, tobacco, cigarettes, aerosols, polishes, hair lotions and lipsticks. It is employed as a soothing ingredient in cosmetic preparations, colognes, deodorants, aftershave lotions and perfume bases. It is also employed in ointments, pain balms, cough syrups, cough lozenges and tablets.

Habitat and distribution

Origin of Japanese mint is not clearly known. It is widely distributed in Brazil, China, Paraguay, Argentina, Japan, Thailand, Angola, India, Bolivia, Peru, Korea and Taiwan. In India, it is chiefly grown in Northern states of Jammu and Kashmir, Punjab and Haryana. World production of Japanese mint oil is about 4000 tonnes/ annum and major producers are Brazil, China and India.

The plant

Among the genus *Mentha*, four major species are commonly cultivated in India.

1. *M. arvensis* L. (Japanese mint)
2. *M. piperita* L. (Pepper mint)
3. *M. spicata* L. (Spear mint)
4. *M. citrata* Ehrh. (Bergamot mint or Lemon mint)

In some cases, Japanese mint is specifically recognized as Menthol mint or *M. arvensis* It is a perennial herb reaching a height of 60-75 cm. It can be grown in subtropical and even tropical areas with cooler climate. It is a perennial herb with rootstock creeping along or just under ground surface. Red, purple and green varieties are known. Branches are rigid and pubescent. Leaves lanceolate to oblong, 3.7-10 cm long, sharply toothed, sessile or shortly petiolated, hairy and have glandular trichomes. Essential oil is synthesised in these trichomes. Flowers arranged in cymes which are usually sessile or rarely pedunculate, purplish, minute; calyx 2.5-3.0 mm long, narrowly deltoid, acuminate; corolla white to purple, 4-5 mm long.

Cultivation

Mints grow well over a wide range of climatic conditions. Japanese mint grows well under subtropical conditions while others prefer temperate climate. Adequate and regular rainfall during growing period and good sunshine during harvesting are ideal for its cultivation. Medium deep soil rich in humus is the best suited for cultivation of Japanese mint. Soil should have a pH range of 6-7.5 with good water holding capacity but water logging is detrimental.

Japanese mint is propagated through stolons. 'CIMAP/MAS-1' and 'Kalka' are improved menthol rich varieties. Recently CIMAP, Lucknow developed some improved cultivars of menthol mint for commercial cultivation. Variety 'Himalaya' is resistant to leaf spot and rest diseases. Variety 'Kosi' is early maturing. Variety 'Saksham' is high yielding and variety 'Kushal' is also early maturing and the best suited for transplanting under delayed conditions. Seed rate is 400 kg/ha. A hectare of well established mint provides enough planting materials for 10 hectares. Mints can be propagated through suckers. Stolons or suckers are planted either on flat land or ridges. In plains, they are planted in shallow furrows of 7-10 cm deep at a spacing of 45-60cm after incorporating compost or farm yard manure at 10-12 tonnes/ha. Inorganic fertilizers up to 160 kg N and 50 kg each of P_2O_5 and K_2O/ha are applied; nitrogen being applied in 2-3 split doses. Irrigation enhances growth and improves yield. For better yield, field requires irrigation at an interval of two weeks. Water logging should be avoided by providing proper drainage. Organic mulch conserves soil moisture and suppresses weed growth. Field should be kept weed free, particularly during initial stages of growth till proper establishment and coverage of ground area.

Japanese mint can be harvested when field is dry. First harvest can be done after about 4 months of planting when lower leaves start turning yellow. Subsequently two more harvests can be taken generally at an interval of 80 days. Yield of herbage from mint crop depends upon climate, soil properties, cultivar used, time of planting and irrigations, control of weeds, pests and diseases, fertilization and proper planting and harvesting schedule. Fresh herb yield ranges from 25 to 50 tonnes/ha and essential oil yield 125-250 kg/ha.

Pest and disease management

Pests

Larvae of mint leaf roller, *Syngamia abruptalis* (Lepidoptera: Pyralidae) roll leaves and feed from within. Insecticides with contact and stomach action keep pest under control. Grubs and adults of red pumpkin beetle, *Aulacophora foevicolis* feed on leaves and buds. Application of mercaptothion at 0.05% or carbaryl 50WP at 4 g/l or DDVP 100EC at 2 ml/l effectively controls pest.

Collar region of plants are cut by cutworm, *Agrotis flammata* and can be controlled by soil drenching with chlorpyriphos 20EC. Hairy caterpillar, *Diacrisia obliqua* feeds on leaves and occasionally causes severe damage. Spraying mercaptothion at 0.05% or carbaryl 50WP at 4 g/l is a remedy against insect. Insect pests such as bug (*Nisia atrovenosa*) and white fly (*Bemisia tabaci*) infest aerial parts of mentha. Insecticides like dimethoate (0.05%), quinalphos (0.05%) or chloropyriphos (0.05%) are effective against these insect pests.

Termites, *Microtermis obesi* and *Odontotermis obesus* attack roots and cause drying of affected plants. Roots are also damaged by white grub (*Holotrichia consaguinea*). Termites are more damaging in summer. FarmYard Manure (FYM), which is not fully decomposed induces termite damage. Chlorpyriphos 20EC at 600-800 ml/ha with irrigation water effectively controls pest.

Root knot nematodes, *Meloidogyne incognita* and *M. javanica* cause yellowing of leaves and stunted growth of plants. Application of neem cake or carbofuran granules @ 20 kg/ha in soil before planting and one year after effectively checks nematodes from damaging levels. Planting of *Tagetes erecta* (marigold) in between plants is also a remedy against nematodes. In highly prone areas, avoid planting for 3-4 years and resort to planting of crops like citronella.

Diseases

Leaf blight: This disease is caused by fungi *Alternaria tenuis* or *Rhizoctonia sp.* Dark brown, circular to irregular patches which coalesce to form bigger necrotic lesions appear by onset of rains in case of *A. tenuis* resulting in defoliation of leaves. *Rhizoctonia sp.* affects nursery plants causing severe loss of planting material. These diseases can be checked significantly by application of mancozeb (0.2%) at fortnightly intervals.

Rust: Mentha rust caused by *Puccinia menthae* results in severe leaf shedding. Rust is the most prevalent disease of mentha throughout world especially in cooler climates of hilly regions. Orange or brick red pustules appear on leaves initially and later lead to defoliation. Application of mancozeb prevents the spread of disease.

Stolon and root rot: Rotting of under ground parts such as stolons and roots is caused by fungi *Macrophomina phaseoli, Thielavia basicola and Rhizoctonia bataticola*. Numerous dark brown necrotic spots appear on stolons. Drenching with mancozeb before planting and pre-treatment of suckers in 0.2-0.3% captan are management measures.

Leaf spot: Corynespora cassicola, Curvularia lunata and *Alternaria spp.* are causal agents of this disease. Necrotic spots appear on leaves and twigs leading to defoliation. Treating the solons with difolatan or spraying with copper oxychloride at 0.3% fortnightly controls disease.

Fusarium wilt: Fusarium oxysporum infection results in yellow, curled and dry leaves. Fortnightly spraying of benlate, carbendazim or topsin at 0.1-0.2% concentration may be resorted to control the disease.

Powdery mildew: Whitish coating and drying of leaves are symptoms of disease caused by *Erysiphae cichoracearum*. Three applications of wettable sulphur at 15 days interval control disease.

Extraction and utilisation

Harvested herb may be wilted in shade for a few hours for draining off excess moisture thereby reducing bulk. Both fresh and dry herb are employed for distillation. To get better result to mint herbage should be shade dried for a day before it is distilled. Steam distillation is usually preferred and duration of distillation is 1.5-2 hours generally. Fresh herb contains 0.4 to 0.6% oil. On an average, 100–150 kg oil/ha is obtained annually. Impurities present in oil can be removed by filtration. Moisture present in oil can be removed by application of anhydrous sodium sulphate. Good quality PVC drum with screw caps are suitable for short term storage of mint oil and galvanized iron drums or aluminium containers are suitable for long term storage.

Physico-chemical properties of oil

Specific gravity at (20°C)	0.8997-0.9011
Optical rotation at (25°C)	-37°11' to 37°29'
Refractive index at (25°C)	1.4590-1.4595
Ester menthol	4.74-5.01%
Total menthol	78.24-82.78%
Acetylated menthol	6.01-6.36%
Menthone	11.85-13.75%
Terpenes	4.1-6.61%
Acid value1	1-2.1
Ester value1	7-18
Solubility in 70% alcohol	2-3 vol.

Chemical constituents in Japanese mint oil are reported to be l-menthone, d and l iso-menthone, methyl acetate, camphene, α-pinene, caryophyllene, esters of formic, iso-valeric and caproic acids, ethyl carbinol, hexanol l-limonene, β-pinene, cineole, 3-octanol, linalool, menthofuran, neo-menthol, pulegone, piperitone and piperitone oxide.

Ocimums

Scientific name: *Ocimum sp.*
Family: Lamiaceae
Vernacular names: *Sanskrit*: Barbari; *Hindi*: Babauitulasi; *Bengali*: Babauitulasi; *Malayalam*: Ramathulasi; *Tamil*: Tirunittur; *Kannada*: Kamakasturi; *Telugu*: Bhutulasi; *Gujarathi*: Damari; *Punjabi*: Babri

Ocimums are important groups of aromatic and medicinal plants which yield many essential oils and aroma chemicals and find diverse uses in perfumery and cosmetic industries as well as in indigenous systems of medicine. Ocimum species with oil rich in camphor, citral, geraniol, linalool, linalyl acetate methyl chavicol, eugenol and thymol are important and can be harnessed for successful utilization by industry. Among various *Ocimum* species, *O. basilicum* L. is commercially and extensively cultivated for essential oil production. Its oil is employed for flavouring of food stuffs, confectionery, condiments and in toiletry products such as mouth washes and dental creams. It is also used in flavouring foods like spiced meats, sausages, tomato pastes, various kinds of sauces, fancy vinegars, pickles, ketchups and beverages. In perfumery industry, oil is used for compounding certain popular perfumes notably jasmine blends. It is recognized as a febrifuge and antimalarial plant. Juice obtained from leaves gives relief to irritation of throat, ear ache and ringworm infections. Seeds are used internally for treatment of constipation and piles.

Habitat and distribution

Ocimums are well represented in warmer parts of both hemispheres from sea level to 1800 m. The main centers of diversity are Africa, South America and Asia. Different species are well distributed over tropical countries in these continents. Of the 160 species of *Ocimum, O. basilicum* is cultivated for at least 3000 years by Europeans and Asians for folklore and religious rituals and got established wherever they migrated with extreme variations in population. It is grown and distilled for oil in France, Italy, Bulgaria, Egypt, Hungary, South America, Comoro Islands, Malagasy Republic, Thailand, India, Haiti and Guatemala.

The plant

Owing to a high degree of polymorphism exhibited by the species and high degree of cross pollination, a large number of species, subspecies, varieties and strains have come into existence which make botanical nomenclature extremely difficult. In view of great diversity, various species are classified into two broad groups, viz. *basilicum* and *sanctum* groups, based on geographical sources, morphological and cytological features and chemical constituents, as detailed below.

Classification of different *Ocimum* species.

Species	**Habit**	**Cytology**	**Major constituents of oil**
Basilicum group (Basic No. x=12)			
O. canum Sims.	Herb	2n=24, 26	Linalool or camphor
O. basilicum L.	Herb	2n=48	Methyl chavicol, methyl cinnamate, eugenol.
O. americanum L.	Herb	2n=72	Methyl chavicol, citral
O. kilimandscharicum Guerke	Shrub	2n=76	Camphor
Sanctum group (Basic No. x=8)			
O. sanctum L.	Shrub	2n=32	Eugenol
O. gratissimum L.	Shrub	2n=40	Eugenol
O. viride Willd.	Shrub	2n=40	Thymol
O. suave Willd.	Shrub	2n=64	Eugenol
O. carnosumck	Shrub	2n=48	Eugenol
O. micranthum	Shrub	2n=48	Elimicin,eugenol,methyl isoeugenol

Additional distinguishing features between two groups are that in basilicum group, bracts are petiolate, flowers more conspicuous and seeds black and ellipsoid which become mucilaginous when wetted whereas in sanctum group, bracts are sessile, flowers less conspicuous and seeds brownish, globose to ellipsoid but do not become mucilaginous when wetted.

Sweet or French basil is an erect, almost glabrous herb, reaching a height of 30-90 cm. Leaves ovate to lanceolate, 3.75-5 cm long; flowers 0.72-1.25 cm long, borne on long terminal racemose inflorescence; calyx 5-toothed; corolla 0.72-1.25 cm long, white to purplish, 2-lipped tube; stamens 4, protruding; ovary bicarpellary, syncarpous, bilocular; stigma bifid; fruit nutlets.

Clocimum is a hybrid strain of *O.gratissimum var. clocimum* obtained by crossing *O. gratissimum* race 1 and *O. gratissimum* race 2, developed at the Regional Research Laboratory, Jammu, India. Name *clocimum* means *clove scented ocimum* and its odour is similar to that of clove oil. Clocimum oil contains 70-75% eugenol and 10-15% myrcene.

Cultivation

Crop comes up well under tropical climate up to an altitude of 1800 m. Growth is poor in areas which receive heavy and continuous rainfall. Frost is harmful to plant and hence frost prone areas are to be avoided. Basil can be cultivated on a wide variety of soils, though moderately fertile, well drained loamy or sandy loam soils with pH ranging from 9.1 to 4.3 are ideal for its cultivation. Basil is tolerant to higher concentration of copper and zinc but is susceptible to cobalt and nickel.

Plant is propagated through seeds. Seedlings are first raised in nursery and then transplanted in field. Seed rate is about 125 g/ha for transplanting. Seeds start germinating 3 days after sowing and germination is over in 7-10 days. When 6-10 cm tall, seedlings are transplanted in field at 40-60 cm spacing in rows. Land should be ploughed well. At planting, 10-15 tonnes of compost or FYM is to be applied. A medium fertilizer dose of 40:40:40 kg/ha of N, P_2O_5 and K_2O is recommended for economic yield though good response is received up to 120:100:100 kg/ha. Irrigation is required once a week when it is raised as a summer crop. Field should be kept weed free for first 20-25 days, till crop canopy completely covers ground. Weeding is usually carried out once or twice.

Basil is harvested when plant is in full bloom and lower leaves start turning yellowish. Leaves and inflorescence are main source of essential oil. Oil content and major oil constituents are maximum during flower initiation and seed setting stage. Crop comes to full bloom 9-12 weeks after planting. For high quality oil, only flowering tops are harvested. Four to five crops are obtained per year. In some areas, it is possible to get four floral harvests. Harvesting is usually done at bright sunny days for high quality oil with maximum yield. First harvest is done when plants are in full bloom and subsequent ones after every 15-20 days. Last harvest comprises whole plant. Floral harvests yield 3-4 tonnes of flowers and an average of 15-20 tonnes herbage can be obtained from each cutting and total yield of 50 tonnes/ha/year. While harvesting whole herb, plants are cut not less than 15 cm from the ground for enabling regeneration of crop.

Pest and disease management

Pests

Bug (*Monanthia globulifera*) causing leaf curling and leaf folder (*Syngamia abruptalis*) feeding on chlorophyll from within leaf folds are major insect pests and can be controlled by spraying of 0.05% phosphamidon 40SL spray.

Nematodes, *Belonolaimus longicaudatus, Meloidigyne incognita, Paratrichodorus christie* and *Pratylenchus leterocephalus* cause either suppression of foliage and root growth or both. Carbofuran 3G or neem cake application in soil is effective against nematodes.

Diseases

Corynespora cassicola causes leaf spot disease, which appears as small water soaked spots turning brown. *Elsinoe arxii* sp. nov. causes scab disease. Symptoms are a little defoliation with pluckering, clipping of leaves and distortion of tender twigs. Blight caused by *Alternaria alternata* and *Colletotrichum capsici* can be controlled by spraying 0.2% zineb or maneb. Wilt is caused by *Fusarium oxysporum* at all stages of growth. But attack is more pronounced in rainy season. Dipping seedlings in a solution of organo mercurial fungicide controls the diseases. Other diseases infecting ocimums are anthracnose (*Colletotrichum gleosporioides, Glomerella cingulata*), blight (*Cercospora ocimicola*), leaf blight (*Pseudomonas syringae*) and collar rot (*Rhizoctonia solanii*). Severe infections by these fungi can be effectively controlled by application of suitable contact fungicides.

Extraction and utilisation

Essential oil in young inflorescence or whole plant is extracted by hydro distillation or steam distillation. Corresponding to part employed, two grades of oil are obtained, ie. flower oil and herb oil. Flower oil has a superior note and is more expensive. Steam distillation is preferred for large plantations as it takes less time and gives better recovery of oil, while hydro distillation carried on in a direct fire still is cheaper and easier for small plantations. Distillation is carried out for 1-1.5 hours.

Essential oil content in whole plant ranges between 0.25-0.47% on fresh weight basis. Young inflorescence contains 0.3-0.5% oil and whole herb 0.10-0.25%. Generally, yields of 30-40 kg flower oil and 20-25 kg whole plant oil are obtained per hectare.

Physico-chemical properties of oil

Refractive index at 24°C	1.528
Optical rotation	6.8°
Specific gravity at 24°C	0.9676
Acid value	2.25

Chemical composition of basil oil is α-pinene 0.1-0.4%, camphene 0.02-0.1%, β-pinene 0.07-0.8%, myrcene 0.12-0.8%, limonene 2.0-9.3%, cis-ocimene 0.1-0.6%, *p*-cymene 0.05-0.15%, cis-3-hexenol 0.02-0.08%, fenchyl acetate 0.1-0.5%, camphor 0.37-0.75%, linalool 40-54%, fenchyl alcohol 2-9%, methyl chavicol 23-26%, α-terpineol 0.8-1.9%, citronellol 0.65-3.7%, geraniol 0.03-0.30%, methyl cinnamate 0.05-0.34% and eugenol 5-12%.

Palmarosa

Scientific name: *Cymbopogon martinii* (Roxb.) Wats.

Family: Graminae (Poaceae)

Vernacular names: *English*: Ginger grass, Rosha grass; *Hindi*: Sofia

Palmarosa is a tall perennial grass, flowering tops and foliage of which contain a sweet-smelling oil of rose-like odour. Two types of palmarosa grasses are recognized, viz. Rosha or Russo grass and Ginger grass. Rosha grass yields a superior oil, used in perfumery, particularly for flavouring tobacco and blending soaps due to lasting rose note that it imparts to blend. In soap perfumes, it has special importance by virtue of geraniol, being stable to alkali. It is a source of very high grade geraniol. Geraniol is highly valued as a perfume and as a starting material for a large number of synthetic aroma chemicals like geranyl esters which have a permanent rose-like odour. It is also used as an adulterant of Turkish attar of roses. Ginger grass oil is poor in quality and is mainly used as a cheap perfume in countries bordering Red Sea.

Habitat and distribution

Exact location of origin of crop is not clearly known. It grows wild in India, particularly in Madhya Pradesh, Maharashtra, Andhra Pradesh, Karnataka, Uttar Pradesh and Orissa. Palmarosa oil is distilled in India for more than 50 years, mostly from wild growths. Plant was successfully introduced into Java, Seychelles, Guatemala and Brazil where commercial production was undertaken to a limited extent. About 60-70 tonnes of palmarosa oil is produced in India annually both from cultivated and natural sources.

The plant

Palmarosa is a perennial grass growing to 2-3 m in height. In this species, following two varieties are recognized which are morphologically similar but differ considerably in their habitat and also in essential oils.

i) C. martinii var. motia (2n=40)

It is known as Palmarosa, Rosha or Russa grass. Leaves are lanceolate, 50 cm long, 1-3 cm broad. Panicles 10-30 cm long, turning reddish, often very bright when mature. Racemes 15-19 mm long, in pairs, each consisting of many pairs of spikelets. In each pair, one is sessile and hermaphrodite and other is pedicelled and male. Sessile spikelets 3.5 mm long. Glumes 3.3-4.0 mm long. Awns 11.4-14.0 mm long. Lower glume of fertile spikelet deeply channeled.

ii) C. martinii var. sofia (2n=20)

It is known as Ginger grass. Morphologically it resembles motia grass, but the oil is inferior. Jamrosa is an interspecific hybrid between C. *nardus var. confertiflorus* and C. *jwarancusa*. Its oil contains 75-80% geraniol and 15-20% geranyl acetate. Oil has more terpenic, limy or grassy by-odours in contrast to sweet, rosy odour of palmarosa oil.

Cultivation

Palmarosa grass is a tropical plant and it grows in warm humid areas. It is susceptible to frost and hence frost-prone areas are not suitable for its cultivation. Although it grows the best on soils having neutral pH, it survives and gives economic yields on alkaline soils of pH upto 9. Motia grass prefers well drained soils and it grows in separate clumps on open dry hill sides with a rainfall of 800-900 mm. It cannot tolerate stagnant water. Sofia grass grows densely and abundantly at lower altitudes in moist and poorly drained soils in areas of higher rainfall.

It is propagated through seeds and also through slips. IW 31245, IW 3629, IW 3244 and Trishna are improved varieties available for cultivation. Seeds are sown on nursery beds prepared in May. About 5 kg seeds are adequate to give seedlings for planting one hectare. As seeds are very small and light, they are usually mixed with fine sand or soil in the ratio of 1:10 for even distribution and ease of sowing. Beds are watered lightly and regularly. Germination starts in two weeks time. In about 4-6 weeks, seedlings are ready for transplanting. Field is prepared well before onset of monsoon and seedlings are transplanted during June-July. Healthy and established seedlings, about 15 cm tall are carefully removed from nursery and transplanted in rows, 20-60 cm apart with plants spaced at 20-60 cm. Spacing can be increased on fertile soils. Farm yard manure is given at 10 tonnes/ha before planting. Fertilizers at 20 kg N, 50 kg P_2O_5 and 40 kg K_2O/ha are given at planting as basal dose. About 40 kg N/ha is applied in two splits during growing season. NPK application should be repeated each year at time of appearance of fresh leaves. Application of micronutrients like iron (as $FeSO_4$) and manganese (as $MnSO_4$) improve growth, herbage and oil yield. Palmarosa plantations are to be irrigated at 10-14 day's interval during summer. Plantation should be kept weed free by regular weeding and hoeing. Diuron at 1.5 kg ai/ha and oxyfluorfen at 0.5 kg ai/ha are recommended to control weeds in palmarosa. Optimum stage of harvest is initial seed setting stage. This stage will be reached about 10-15 days after flowering. Grass is cut at a height of about 10 cm from ground level and whole plant is used for distillation. During first year one or two cuttings can be obtained depending upon climatic conditions. After first harvest, subsequent harvests can be made at 70-80 days interval and 3-4 cuttings can be taken a year. Plantation remains productive for 4-6 years. Yield of grass and oil starts decreasing from third or fourth year onwards. Grass yield is 6-10 tonnes/cut/ha.

Pest and disease management

Pests

Aphids: Adults and nymphs of *Aphis gossypii* (Hemiptera: Aphididae) suck sap from inflorescence of plants. Attack is maximum during summer from January-April. Spraying dimethoate at 0.3-0.7 kg a.i/ha is effective to control pest.

Thrips: Adults and nymphs of yellowish brown, 1.5 mm long thrips namely *Haplothrips sp.* (Thysanoptera: Thripidae) damage young shoot tips and leaves and also feed on floral parts affecting seed setting. Attack is more severe during Februvary-April and July-August. Dimethoate spraying will suppress pest damage.

White grub: Grubs of *Holotrichia consanguinea* (Coleoptera: Melolonthidae) feed on roots of palmarosa. Grub is dirty white or brown coloured and severe damage is during June-November. Chlorpyriphos 20EC at 600-800 ml/ha with irrigation water effectively controls pest damage.

Termite: Microtermis sp. (Isoptera: Termitidae) is the major termite, attacking palmarosa. Whitish coloured adults damage crop throughout year. Newly planted seedlings are more vulnerable and plants die when basal portion of stem is eaten away. Chlorpyriphos 20EC with irrigation water controls pest.

Diseases

Palmarosa cultivation in large scale sometimes attracts two major fungal diseases viz. Ellisiella blight and Curvularia blotch.

Ellisiella blight: It is one of the serious diseases of Palmarosa caused by *Ellisiella caudate.* This disease appears in epiphytotic form during rainy season and causes considerable amount of loss in production of herb and essential oil. Small grey necrotic spots appear as initial symptom on the surface of infected leaves. In severe cases lesions get enlarged and coalesce resulting in premature drying of infected leaves. Fungus produces masses of spores on dried necrotic lesions. Disease can be effectively controlled by foliar spraying captafol or chlorothalonil @ 0.3% at 15 days interval.

Curvularia blotch: This disease is prevalent in Palmarosa growing areas of U.P, M.P, Bihar, Karnataka and J&K and is caused by *Curvularia andrographis* and *C. trifolii*. Disease occurs in epiphytotic form during August and October. Small eye shaped, orange/brick red necrotic lesions appear and coalesce together resulting in premature drying of leaves. Foliar application of mancozeb @0.3% at 15 days interval at initial stages of infection effectively controls the disease.

Dreschlera cymmartinii also infects palmarosa. Application of a suitable fungicide can be resorted to, if attack is severe.

Extraction and utilisation

Harvested herbage is allowed to wilt in shade for 24-48 hours for draining off excess moisture from leaves. This reduces bulk and cost of distillation. Oil can be obtained either by hydrodistillation or by steam distillation. Steam distillation yields more of better quality oil. Distillation unit should be clean, rust free and free of any other odour. Oil content and yield depend upon climatic conditions, harvesting time, maturity of grass, extent of wilting and distillation process. Oil yield is low in first year and it increases with age but gradually decreases after fourth year. All parts of plant contain essential oil, the maximum being present in flowers and the least in stalks. On an average, oil content in various parts are: whole plant 0.10-0.40%, stalks 0.01-0.03%, flowering tops 0.45-0.52% and leaves 0.16-0.25%. Average annual oil yield is 100-125 kg/ha though a yield of 250 kg/ha is not uncommon.

Physico-chemical properties of oil

Physico-chemical properties of the two oils are as follows.

Property	Rosha grass oil	Ginger grass oil
Specific gravity (15°C)	0.887-0.900	0.900-0.955
Optical rotation	-2° to 3°	-30° to 54°
Refractive index (20°C)	1.468-1.476	1.479-1.493
Acid value	upto 1.8	upto 2
Ester value after acetylation	9-36	120-200
Total alcohols calculated as geraniol	80-95%	36.3-64.7%
Solubility in 70% alcohol	1-3 vol.	2-3 vol.

Pale yellow rosha grass oil has a sweet, floral rosy odour and various undertones and top notes according to quality and age of oil. Yellowish-brown ginger grass oil has a peculiar fatty-sweet odour leaving a slightly woody and rosy dry out note, sometimes referred to as an "ensilage" odour.

Best natural source of geraniol is motia oil containing up to 95% geraniol. Chemical composition of motia oil is reported as limonene 0.1%, *p*-cymene 0.1%, methyl heptanone 0.1%, 2-nonanol 0.1%, linalool 2.4%, citronellol 6.4%, farnesene 0.6%, β-terpineol 1.0%, β-humulene 0.6%, α-terpineol 0.4%, geraniol 81.7%, geranyl acetate 5.7% and farnesol 0.4%.

Patchouli

Scientific name: *Pogostemon patchouli* Pellet var. *suavis* Hook.f; Syn: *P. cablin* Benth
Family: Labiatae
Vernacular names: *Sanskrit*: Pachi; *Hindi*: Pachauli; *Bengali*: Pachapat; Patchouli; *Malayalam*: Pachila, Kattam; *Tamil*: Kadir pachai, Kattam; *Kannada*: Pachetene; *Gujarati*: Pacha, Sugandhi pandi; *Marathi*: Panch

Patchouli leaf oil is used as a perfume by itself and it is one of the best fixatives used for heavy type perfumes and it is also used in soaps, cosmetics and flavour industries. It is used as a base note. Oil is extensively used as a flavour ingredient in major food products, including alcoholic and non-alcoholic beverages, frozen dairy desserts, candy, packed foods, gelatin, meat and meat products. It blends well with oils of sandal wood, geranium, vetiver, palmarosa, cedarwood, oakmoss, clove, cassia, rose, lavender and bergamot. Oil gives one of the finest attars when blended with sandal wood oil. Tenacity of odour is one of the great virtues of patchouli oil and is one of the reasons for its versatile use.

It is used in Asian medicine as an aphrodisiac, antidepressant and antiseptic. Oil possesses anti bacterial, antifungal, antiviral, antiinflammatory and insect repellent activities. It reduces blood congestion. It is used in anti wrinkle creams. It is used as a moth repellent in India. In Chinese medicine, patchouli leaves are used as decoction with other drugs to treat nausea, diarrhoea, cold and head aches. Dried leaves are used for scenting wardrobes. Leaves and tops are added in bath water for their antirheumatic action. It is also used as a masking agent for alcoholic breath. In aromatherapy it is used as a relaxant. It can ease and diminish anxiety and depression. In high dose, it stimulates and in lower dose, it is sedative. It is used to improve intelligence and concentration of mind.

Habitat and distribution

Patchouli is a native of Malaysia and Philippines. It is widely cultivated in Malaya, Sumatra, Java, Seychelles, Madagascar, Reunion Island, Paragway and Sao Paulo. It grows wild in Malaysia, Indonesia, Singapore, China and India where they are also cultivated for essential oil. It was introduced to India in 1942. In India, it is seen in Southern Peninsula, notably in Western Ghats, Nilgiris, Central India and subtropical Himalayas. World production of patchouli oil is around 2000 tonnes/annum and Indonesia meets 90% of world requirement of oil. There is no synthetic chemical to replace oil of patchouli. Domestic production of patchouli is around 250 tonnes/annum and most of domestic requirement is met by import from Indonesia, Malaysia and Singapore.

The plant

Patchouli is a herbaceous, erect, branched, pubescent aromatic herb, 0.5-1.0 m high. It has square stem swollen on nodes, oval leaves, about 10 cm long and wide, serrated with dotted glands beneath and about 8 cm long petiole. Spikes terminal or axillary, dense, sometimes interrupted, 8-15 cm long. It flowers reluctantly and flowers are white to light purple in colour, very small, irregular, bisexual, hypogynous; calyx 5-6.5 mm; corolla lobes obtuse, 6-9 mm, white to purple, blotched on all segments; filaments violet; bracts as long as calyx.

There are a few related *Pogostemon* species. *P. heyneanus* Benth. known as Java patchouli flowers and leaves are thinner and not lobbed in upper part and faintly serrate. It is indigenous to India. Oil produced has very inferior odour. *P. paniculatus* Benth. is grown near rocky slopes in Kerala and Maharashtra. *P. plectranthoides* Desf. is more common in Indian forests and yields 0.5% of thick viscous oil of not very pleasant odour.

Cultivation

Patchouli prefers warm humid climate with an evenly distributed rainfall of 250-300 cm/ annum, a temperature of 24-28°C and an average atmospheric humidity of 75%. It grows up to an altitude of 1000 m above MSL. In low rainfall areas, crop can be grown under irrigation. It is relatively a hardy plant adapted to a wide range of soil conditions. A well-drained deep loamy soil rich in humus and nutrients with a loose friable structure is ideal. A pH range of 5.5-6.2 is suitable. Patchouli is a shade loving plant and comes up well in partially shaded conditions also. It is generally grown as an intercrop in crops like papaya, coconut or arecanut.

Improved types commonly cultivated are Johore, Singapore and Indonesia. Variety IIHR 5 showed good response to nitrogen fertilizer application. Plant is propagated by 4-5 noded stem cuttings of 15-20 cm length. Terminal cuttings after removal of basal 2-3 pairs of leaves are used. Treatment with IBA, IAA or NAA at concentrations of 500, 1000 or 1500 ppm respectively is found to promote rooting. Cuttings are planted at 3-5 cm apart in nursery beds, seed pans or polythene bags and partial shade and regular watering are provided. Rooting occurs in 4-5 weeks and they are ready for transplanting in 8-10 weeks. Mass propagation technique for patchouli through somatic organogenesis using leaf and stem as explants has also been standardized.

For field planting, beds of convenient size are made and organic manure at 12-15 tonnes/ha and N, P_2O5 and K_2O at 25:50:50 kg/ha are incorporated and levelled. Rooted cuttings are transplanted at 60x60 cm spacing. Adequate soil moisture should be ensured during initial stages of field establishment. A wider spacing of 90 cm is superior in terms of herbage and essential oil yield. Top dressing with 25 kg N/ha is to given thrice, two months after planting, just after harvest and then two months later. Herbage, oil yield and oil quality of patchouli are influenced by irrigation, organic mulch and nitrogen application in semi-arid tropical climate. Regular watering, timely weeding and light cultivation after every harvest are essential for satisfactory crop performance.

Crop is harvested when foliage becomes pale green to light brown and emits a characteristic patchouli odour. Shoots of 25-50 cm length which contain at least 3 pairs of mature leaves are harvested. Oil yield from mature leaves is higher than that from tender leaves. First harvesting can be done after five months of planting and subsequent harvests at 3-4 months interval. A few shoots are left unplucked to ensure better growth for next harvest. Harvesting should be done in cool hours of morning to avoid loss of essential oil. Crop can be retained for 3-4 years after which new planting is required.

Pest and disease management

Pests

Patchouli is severely infested by root knot nematode, *Meloidogyne incognita* which causes gall swellings on roots. Infected plants show water stress and stunted growth due to lack of nutrients. Number of leaves gets reduced in affected plants and turn pale green. Phytosanitory measures are to be adopted from nursery stage itself. An integrated approach consisting of crop rotation, application of neem oil cake, nematicides and carbofuran are effective in controlling nematode problem. Nematode damage is effectively managed by application of carbofuran 3G @1.0 kg a.i/ha in mainfield and nursery.

In south India, leaf feeding caterpillar (*Pronomis profusalis*), leaf webber (*Pachysancla aegrotalis*) and leaf roller (*Herpetograma licersisalis*) are major insect pests. Bug, *Pachypeltis sp*. feeds sap from leaves. If infestation is severe, field should be sprayed with mercaptothion (0.05%) or methyl parathion (0.05%) at fortnightly intervals. Carbaryl 50WP at 4 g/l is also effective.

Diseases

Root rot and wilt caused by *R. solani* and *F. oxysporum* are major diseases of patchouli, especially when soil moisture is high. Providing proper drainage and soil incorporation of Pongamia cake @ 1.0 tonne/ha before plating are remedial measures of disease.

Leaf blight caused by *Cercospora sp.* is controlled by spraying zineb 0.5%. Yellow mosaic disease, transmitted by white fly, *Bemisia tabacci* (Gen.) can be effectively managed by controlling vector.

Yellow mosaic of leaves caused by virus Potyviridae is transmitted by aphids. Chlorotic spots which later coalesce together to form larger mosaic areas are more prominent in summer. Application of dimethoate @ 0.05 % checks insect vector and thereby virus transmission.

Patchouli mild mosaic virus (PaMMV) decreases leaf biomass and essential oil yield. Transgenic patchouli plants with PaMMV coat protein precursor gene were produced by agrobacterium-mediated transformation by Tokyo University of Agriculture, Japan. Transgenic patchouli plant with PaMMV CP-P gene is highly resistant to PaMMV. During 1996-97, a new phytoplasma disease causing witches' broom in patchouli was reported from India. Characteristic symptoms include leaf shrinkage, axillary proliferation of branches, shortened internodes and stunted growth and bushy appearance of plants. Flowers exhibited phyllody symptoms. Application of Oxytetracycline hydrochloride solution as foliar sprays or soil drench is effective in checking disease.

Extraction and utilisation

Harvested herb is dried in shade for about 3 days. During drying, material should be frequently turned over for promoting uniform drying and for preventing fermentation. Completely dried material can be pressed into bales and stored in a cool dry place for a short period, if required. In case of fresh materials, distillation for three hours is sufficient whereas in case of semidry leave, 5-6 hrs and for dry leaves, 7- 9 hrs are required. Shade drying and storage for 150 days are congenial for maximum recovery of oil.

Prolonged distillation gives higher yield and better quality of oil. But if it is distilled for too long, oil will have a disagreeable odour. The aged has a finer aroma than the fresh ones. Interchange of high and low pressures (1.4 to 3.5 kg/cm^2) produces better yield as more cell walls rupture in this process.

Oil yield varies from 2.5 to 3.5% on shade dry basis. On an average, crop yields 8000 kg/ha fresh leaves which on shade drying yields 1600 kg dry herbage which in turn gives 40 kg of oil on distillation. Patchouli resinoid can be prepared by extracting leaves with volatile solvents such as benzene. Solvent extraction yields 4.5-5.8% of resinoid containing 70-80% of alcohol soluble absolute.

Physico-chemical properties of oil

Patchouli oil of good quality is characterized by high specific gravity, high laevo rotation, high refractive index and good solubility in 90% alcohol.

Essential oil of patchouli has the following properties.

Specific gravity at 15°C	0.967-0.972
Refractive index at 20°C	1.509-1.510
Optical rotation	49° 40'-55° 41'
Acid value	about 5.0
Saponification value	5.6 - 10.7
Ester value after acetylation	16.8-21.5
Solubility in 90% alcohol	6.5-7.0 vol.

Newly distilled patchouli oil has a fresh green slightly harsh aroma. Its odour improves on aging, becomes sweeter and balsamic if stored well for some months. Oil matured for several years possesses a fine and fuller odour than fresh oil and is highly esteemed by perfumers. Patchouli oil has following approximate composition: caryophyllene 20%, α-bulnesene 25%, α-guaiene 15%, patchouli alcohol 30%, α-bulnesene oxide 4%, caryophyllene oxide 2%, pogostol 1%, b-elemene 1%, α-guaiene oxide 1%, nor-patchoulinol 0.5%, α-patchoulene, β-patchoulene, δ-patchoulene and seychellene. The most odour intensive constituents of patchouli are patchouli alcohol and nor-patchoulinol.

Sandalwood

Scientific name: *Santalum album*
Family: Santalaceae
Vernacular names: *English*: Sandalwood tree; *Bengali*: Chandan, Sadachandan, Pit chandan; *Gujarati*: Sukhada; *Hindi*: Safed chandan, Safed sandal; *Sanskrit*: Srigandha, Swetha chandan, Chandanam, Gandhshrah; *Telugu*: Gandhapuchekka, Srigandhapumanu; *Tamil*: Shadanalekattai, Shandanamaram, Chandana-kattai; *Malayalam*: Chandanamaram

The term sandalwood, in the world market, is frequently used for a variety of woods that yield oils similar in smell to that of the East Indian Sandalwood which is the true sandalwood. The world famous East Indian Sandalwood oil is extracted from the strongly scented heartwood of this tree. Sandalwood oil is used primarily in perfumery because of its outstanding fixative properties. It is used in preparing all types of perfume compositions especially Indian attars like Hina, Gulab, Kewda and Jasmine in which the natural essential oils from distillate of floral distillation is absorbed in sandalwood oil. With neem oil, it is used as contraceptive. It is used for healing wounds and blisters caused by the smallpox vaccination. Sandalwood is also one of the finest woods for carving. Wood is smooth with uniform fibres. Saw dust from heartwood is mostly used in incense for scenting cloths and cupboards.

Habitat and distribution

Sandalwood tree is indigenous to mountain districts of south India and Malayan Archipelago. Plant historians believe that tree is indigenous to south East Asia (Timor Islands) and was introduced into India by traders possibly before Christian era. In India, sandal is spread over 480 km from Dharwar in north to Nilgiris in south and 400 km from Coorg in west to Kuppam (Andhra Pradesh) in east. About 90% of world production of sandalwood oil is from India.

The plant

Sandalwood is a small evergreen tree growing to 18 m in height and 2.4 m in girth, with slender drooping branches. Sapwood is white and odourless while heartwood is yellowish brown and strongly scented. Leaves 3.5-4 cm in length, elliptic, lanceolate glabrous and petiolate; inflorescence terminal or axillary, paniculate cyme; flowers bisexual, many, brownish purple; perianth campanulate; stamens 4, exerted, alternating with 4 rounded obtuse scales; fruit drupe, globose, 1.25 cm in diameter, purple black, endocarp hard and ribbed. Sandal tree is a plant parasite and its roots thrive on many types of host plants such as *Cassia siamea, Pongamia pinnata, Lantana acuminata* and *Cajanus cajan*.

Cultivation

Warm tropical climate is the best suited for sandalwood tree. It grows well between altitudes of 600 m and 1350 m above MSL though it grows between 360 m and 1850 m altitude. Annual rainfall of 600-1600 mm is ideal for its growth. More than 1800 mm of rain is not very conducive to its growth. It grows well on laterite soils on slopes of hills exposed to sun. Plant is propagated through seeds. Barring a few tissue culture attempts, vegetative propagation has not been very successful. Seeds are obtained from plants over 20 years old. Fresh seeds obtained from October fruiting are depulped, dried and sown on seed beds. Gibberellic acid is used to bring down dormancy period and to induce quick and uniform germination. After germination, seeds are put in polybags of size 15 cmx25 cm. A host plant is sown in polybag when seedling reaches 15 cm in height. Optimum stage for planting is when seedlings are 25-50 cm high and basal portion becomes darker. Pits of 30-50 cm cube are dug and sandal seedlings along with host seedlings are planted from May to October at 2.5-4.0 m spacing. Weeds are removed as and when necessary. In case, sandalwood seedlings are overtopped by host plant, the host is lopped to provide sufficient light to seedlings. Heartwood formation is at its peak when trees are 30-60 years and trees attain a girth of 40-60 cm. Sandal spike disease is caused by mycoplasma-like organisms which cause severe reduction in leaf size and shorten internodes. As roots are the richest in oil, sandalwood tree is harvested by uprooting and not by cutting to avoid loss of root system.

Extraction and utilisation

Sandalwood oil is obtained chiefly by steam distillation of powdered wood soaked in water for about 48 hours. Distillation is carried out at a steam pressure of 1.4-2.8 kg/cm^2 for 48-72 hours. Oil content is about 10% in roots and 1.5-2% in chips which constitute a mixture of heartwood and sapwood. Yield from heartwood varies with maturity and locality.

Physico-chemical properties of oil

Sandalwood oil has a persistent woody odour and following properties.

Specific gravity (20 °C)	0.971-0.983
Optical rotation (20 °C)	-15° 58' to -20°
Refractive index (20 °C)	1.505 to 1.510
Alcohol calculated as santalol	> 90%
Acetate calculated as santalyl acetate	> 20%
Solubility in 90% alcohol	> 1 vol.
Solubility in 70% alcohol	5 vol.

Major constituents of sandalwood oil are α-santalol 60%, β-santalol 30%, α-and β-santalene, α- and β-curcumene, β-farnesene, santene, santenol, santenone, teresantalol, teresantalic acid, santalic acid, nor-tricyclockasantalol, borneol and isovaleraldehyde.

Vetiver

Scientific name: *Vetiveria zizanioides* (Linn.) Nash; Syn: *V. odorata* Virey, *Andropogon muricatus*
Family: Graminae (Poaceae).
Vernacular names: *Sanskrit*: Usirah, Vira; *Hindi*: Khas; Bengali: Khas-khas; *Malayalam*: Ramacham; *Tamil*: Illamichamber; *Kannada*: Vattiveru; *Telugu*: Vattiveru; *Gujarathi*: Valo; *Punjabi*: Panni; *Marathi*: Vala; *Urdu*: Khas

Vetiver roots contain a fragrant oil, considered as one of the finest aromatic oils. It is a perfume by itself. Aroma chemicals such as vetiverol, vetiverone and vetiveryl acetate are prepared from this volatile oil. Oil is a high grade fixative and blends well with sandal wood, lavender, patchouli and rose oil and used in perfumes, soaps, skin care products etc. In India, it is used in tobacco and pan masala industries. Roots are often kept along with clothes to repel insects. Root is cooling, bitter, alexiteric, stomachic, carminative, astringent, stimulates immune system, promotes menstruation, useful in headache, burning sensations, ulcers, rheumatism and diseases of blood. Vetiver oil is used in snake bites, cancer and microbial infections.

Since plant has extensive fibrous roots, it is useful in both soil and water conservation. It helps in maintaining soil moisture, absorbs toxic substances in chemical fertilizers and pesticides and improves physical characteristics of soil. Dry roots are used for making mats, fans, screens, pillows, baskets, incense sticks and sachet bags. Roots after oil extraction are used as a raw material for making cardboard, paper etc.

Young leaves are used as fodder and also as bedding material for horses and cattle. Dry leaves are used for thatching purposes and for making brooms. Leaves made into pulp are suitable for making straw boards. The above ground portion is used in various ways such as making paper, ropes, mats, hats, baskets, for mulching, as substrate for mushroom culture and for making compost.

The plant is one of the best soil binders and is used throughout tropics to check soil erosion by planting along contour. Vetiver with its strong roots and a rooting depth of about 5 m, was identified by World Bank as most promising green technology against erosion. Grass is widely grown as protective partitions in terraced fields and as a border for roads and gardens. About 10 year-old vetiver is the most efficient barrier to reduce soil, nutrients, organic matter and water losses. It can effectively check leachate in municipal and industrial waste dumps and reduce silting up of rivers and dams. It accumulates cadmium, lead, mercury, arsenic and nickel and tolerates many organic poisons making it suitable for phyto- remediation of heavy metal contaminated soil. It is one of the best species for re-vegetation of mine tailings containing high levels of heavy metals (Pb, Zn, Cu, and Cd) and low levels of major nutrient elements (N, P, and K) and organic matter. Heavy metal toxicity and extreme infertility are major constraints on re-vegetation in such areas.

Vetiver oil is a highly effective repellent against the Formosan subterranean termite, *Coptotermes formosanus* because of its long-lasting activity. It decreases termite tunneling activity and food consumption. Vetiver oil disrupts termite behaviour and physiology as a consequence of direct physical contact, ingestion, or exposure to vapours. Ingestion of wood treated with vetiver oil results in progressive death of protozoa population living inside termite gut. It is a promising novel termiticide with reduced environmental impact.

Habitat and distribution

Vetiver is indigenous to India, Pakistan, Bangladesh, Sri Lanka and Malaysia. Haiti, Indonesia, Guatemala, India, China and Brazil are the main producers. World production of essential oil is around 250-300 tonnes/annum. Haiti and Indonesia account for 80 % of total vetiver oil production in world. Crop is also cultivated in Indonesia, Malaysia, Philippines, Japan, Angola, Belgian Congo, Dominican Republic, Argentina, British Guiana, Jamaica, Mauritius and Honduras. In India, it is seen growing wild throughout Punjab, Uttar Pradesh and Assam. It is cultivated in states of Rajastan, Uttar Pradesh, Karnataka, Tamil Nadu, Kerala and Andhra Pradesh. About 20-25 tonnes oil is produced in India annually, much below its demand. Uttar Pradesh produces the highest quantity of oil, mainly from wild plants. Vetiver oil produced from North India is the best and costliest in the world. Though, it grows well on hillsides, natural habitat is low, damp sites such as swamps and bogs. Plant is commonly found along the waterways.

The plant

There are seeding and non seeding cultivars. North Indian types are usually seeding and generally possess better aroma. Its rooting tends to be shallow, especially in damp ground. South Indian types are mostly non-seeding. Yield of South Indian type is higher and it is a densely tufted, wiry, glabrous perennial grass with a thicker stem, with less branching roots and wider leaves and it is the cultivated type. This is distributed throughout tropics and probably a selection from wild type. It is non- flowering, non- spreading, reproduced by vegetative propagation and it is the type suitable for erosion control.

Vetiver is a diploid with a chromosome number 2n=20. It grows in large clumps from a much branched spongy rootstock with erect culms, 1-2 m high usually sheathed all along. Leaf sheaths are compressed, imbricate, very smooth and firm. Leaf blades are stiff, long, narrow, and 0.75-1 m in length, 8 mm or less in width, glabrous, but rough on edges. Leaves are odourless and do not contain essential oil. Roots are strongly scented and are of 10- 35 cm length.

Many cultivated types rarely flower. Flowers that are seen in some, seldom set seeds. Panicles are 15-30 cm long, very narrow; branches 2.5-5.0 cm long, whorled; spikelets in pairs, narrow, acute, appressed, awnless; one sessile and hermaphrodite, somewhat flattened laterally, with short sharp spines, 3 stamens and 2 plumose stigmas; the other spikelet pedicelled and staminate.

Cultivation

Vetiver prefers warm humid tropical and subtropical climate. It grows in areas with an annual rainfall of 600-2000 mm with moderately humid climate, up to an altitude of 1000 m, in a temperature range of 21°C to 44°C. It is mainly cultivated as a rainfed crop in hill slopes. It can also be grown as an irrigated crop in places of scanty rainfall. Plant requires plenty of sunlight and long day condition. Vetiver grows on almost all types of soils but a rich and well drained sandy loam is the best. It grows in saline and sodic soils, sandy soils, riverine soils, marshy areas and tolerates high degree of water logging and moisture stress. Roots from crops grown in light soils yield very low percentage of oil whereas roots obtained from red lateritic soils with abundant organic matter are thick and contain more essential oil. Heavy soils make harvesting of roots difficult, with a loss of finer roots which contain most of oil.

In India, two types, namely South Indian and North Indian are generally under cultivation. North Indian types yield oil of superior quality whereas South Indian types are high yielders both in terms of root mass and oil. 'Pusa hybrid-7', 'Hybrid-8', 'CIMAP/KS-2', 'Sugantha' KH-8, KH 40 and 'ODV-3' (South Indian type) are improved varieties available for commercial cultivation. Nilambur type (ODV- 3), on an average produces 5 tonnes/ha of root and 20-30 kg oil/ha.

Cultivars Dharini, Gulabi and Khesari released by CIMAP Lucknow were developed by repeated selection from germplasm collections from different parts of India. They produce high essential oil yields but aromas of oils differ. Gulabi and Dharini are tolerant to sodic soils. Dharini possesses khus note. Dharini has longer and thicker roots and is a good soil binder and useful for soil and water conservation. Gulabi yields about 2.8 tonnes of dry roots and 25-30 kg essential oil from one hectare. It can also be cultivated in marginal soils and wastelands. The best planting season for this variety is June-July. It has a rosy odour whereas Khesari has saffron odour.

The grass can be propagated either through seeds or slips, but slips are commonly used. The cultivated accessions propagated through vegetative means show limited variation. Seed propagation is limited to breeding of new varieties. In North Indian types, profuse seeding occurs and natural regeneration occurs from self sown seeds. Pure germinating seed yield varies between 400-650 kg/ha. Freshly collected spikelets show dormancy and require an after-ripening period of about 3 months. Removal of caryopsis from enclosing husk facilitates germination. Dormancy can also be broken by treating with gibberellic acid or potassium nitrate.

In South Indian types, most spikelets are not subject to fertilization and seeds, which are sometimes produced are very thin and have a short dormancy period. In these non-seeding types, slips separated from clumps of previous crops with rhizome portion intact having 15-20 cm of aerial portion is used. Slips thus obtained should be kept moist and stored in shade. Dry leaves are removed from slips to avoid chances of spread of diseases. The most suitable time for planting is early June to early August with onset of monsoon. In South Indian conditions, where diurnal variation in temperature is not significant and monsoon sets in early, optimum planting time is February to April. While planting slips, fibrous roots and leaves should be trimmed off. With onset of monsoon, land is prepared by 2-3 deep ploughings and after removing perennial weeds, farm yard manure or compost is applied at 5 tonnes/ha. In sloppy areas, pits are taken across contour. Slips from healthy and disease free clumps are planted during June-July with onset of monsoon vertically about 10 cm deep at a spacing of 60x30 cm, 60x45 cm or 60x60 cm based on soil fertility status, climate, variety and irrigation facility. Plant population varies from 27800 to 1,10,000 plants/ha. Late planted crop yields coarse roots and inferior quality oil.

Normally, fertilizer application is not practised in fertile soils. On poor soils, N, P_2O_5 and K_2O may be applied each at 25-50 kg/ha of which N may be applied in 2-3 split doses. Application of 60 kg P_2O_5/ha is suggested for vetiver cultivation in central Uttar Pradesh. From planting to sprouting, soil moisture status should be maintained by irrigating in absence of rainfall. In areas with well distributed rainfall and high humidity, irrigation is not necessary and in other areas, 8-10 irrigations are required. Two to three weedings and earthing up at an interval of one month are needed during initial period of plant growth. Once plantation is established very well, weeds are kept under check because of thick and dense shoot cover. Aerial portion is trimmed about 20-30 cm above ground thrice during period of two years, first trimming at 4-5 months after planting and second during second year just before flowering and third in second winter, about one month before digging of roots. When vetiver grass is grown for phyto-remediation of heavy metal contaminated soil, above ground biomass should be regularly cut to stimulate re-growth and translocation of heavy metals to shoots.

In many tropical countries, vetiver grows and survives without nitrogen and phosphorus fertilizer application because of associative nitrogen fixation. Heterotrophic microbes associated with vetiver root are nitrogen-fixing bacteria, phosphate- solubilizing microbes, mycorrhizal fungi and cellulolytic microorganisms. Most of N-fixing bacteria are present on root surface or in intercellular spaces or in dead cells within the root. Many plant hormone derivatives are produced by N-fixing bacteria such as *Azotobacter*, *Azospirillum*, *Bacillus* and *Pseudomonas*. Several soil bacteria, particularly those belonging to genera *Pseudomonas* and *Bacillus* possess ability to change insoluble phosphates in soil into soluble form by secreting organic acids such as formic, acetic, propionic, lactic, glycolic, fumaric and succinic acids. These acids lower pH and bring about dissolution of bound phosphates. Bacterial inoculation of vetiver grass induces branching of fibrous roots and increases plant dry weight.

Fungi such as *Penicillium* and *Aspergillus*, can change insoluble phosphates in soil into soluble form that affects plant growth. Ectomycorrhiza and endomycorrhiza increase surface area of root system for better absorption of nutrients from soil, especially when soil is deficient in phosphorus. Endomycorrhiza, known as vesicular-arbuscular mycorrhiza (VAM), possesses special structures that help in transfer of nutrient from soil into root system. VAM associations especially that of *Glomus mosseae* improve uptake of macro and micro-nutrients and stress tolerance.

Strongly positive correlations are reported between plant height, root length and oil yield; root length and oil content; fresh and dry root yield and oil yield and oil content and oil yield. These traits form a good selection criteria for improvement of essential oil yield in vetiver.

Roots are harvested after 15-24 months of planting but for maximum oil yield of good quality at 18 months. Earlier harvesting gives higher yield of oil, but of low specific gravity and lacking in the valuable high boiling constituents. If roots stay in ground for over two years, oil quality improves but yield diminishes considerably and oil becomes very viscous with a dark colour. Crop is generally harvested during December-February by digging out clump along with its roots. A tractor drawn mould board plough can be used for digging out roots up to 35 cm depth. Mechanical harvesting gives 15% higher roots recovery over manual digging. Length of roots varies from 10-35 cm. Roots are washed and dried under shade for 1-2 days which improves olfactory quality of essential oil. Prolonged drying in sun reduces oil yield. On an average, root yield is 3-4 tonnes/ha from a two year old plantation. In sandy and sandy loam soils, root yield is about 2-2.5 tonnes/ha and in salt affected areas about 1-1.5 tonnes/ha.

Pest and disease management

Vetiver is a very hardy crop. Infestation by diseases and pests is not of serious concern. In dry areas termites are seen damaging vetiver. Grubs of beetle *Phyllophaga serrata* have also been reported infesting vetiver roots. These can be controlled by broadcasting lindane dust at 25 kg/ha before final ploughing. Stem borer, *Chilo sp.* is also a threat to commercially grown vetiver.

Fusarium blight is seen during rainy season. Leaf blight disease caused by *Curvularia trifolii* is another important disease in rainy season.

Leaf blight caused by *Curvularia trifolii* and *Fusarium* disease is controlled by repeated spraying and drenching with copper oxychloride or 1% Bordeaux Mixture. Scale insects are kept under check by the application of metasystox (0.4%) or chlorpyriphos at 2.5 l/ha. Root infesting beetles *Phyllophaga serrata* is controlled by broadcasting lindane dust at 25 kg/ha before final ploughing.

Extraction and utilisation

Roots are crushed, powdered and oil is extracted through hydro or steam distillation. Both fresh and dry roots can be distilled. Roots are shade dried and chopped into small pieces. About 15-16 hours are required for distillation. Generally, the South Indian varieties require a longer time. Its oil has low volatility and high boiling point. Two distinct fractions one lighter than water and other heavier than water are obtained from vetiver. Heavier the oil better the quality. These fractions should be collected separately and later mixed together. Water content is totally removed by exposing to sun in an open container or by using a centrifuge. Roots must be distilled for a prolonged period ranging from 24 to 48 hours since the most valuable quality constituents are contained in high boiling fractions. Oil obtained from stored roots is more viscous and possesses a slightly better aroma than that obtained from freshly harvested roots. To obtain maximum oil yield and to shorten time of distillation, roots should be distilled when fresh. Oil recovery from fresh roots is 0.3-0.8% and from dried roots is 0.5-3.0% depending up on duration of distillation. On an average, oil recovery is around 1% only on d. w.b and 10-30 kg oil is obtained per hectare per crop.

Physico-chemical properties of oil

Vetiver oil is light brown to deep brown in colour with a characteristic aroma and persistent odour of sweet woody note. Aging for a period of six months improves the odour of oil substantially. The harsh, green and earthy odour of freshly distilled oil disappears and develops into a fuller, heavier and sweeter odour.

Two distinct types of oils, highly laevorotatory oil from wild North Indian types and dextro-rotatory oil from cultivated types grown both in South India and outside India are obtained. North Indian oil contains sesquiterpenes such as khusol, khusonol, khusitone and leavojuneol etc. North Indian variety of vetiver oil contains not less than 70 % vetiverol and total ketone content not less than 24%. Presence of more of hydrocarbons and less of oxygenated constituents in oils from South Indian varieties contribute to its low olfactory value.

Other properties are:

Parameters	South Indian	North Indian
Specific gravity at 30°C	0.990 - 1.015	0.991 – 1.033
Refractive index at 30°C	1.516 - 1.530	1.512 – 1.523
Optical rotation	+10° to +25°	-50° to -132°
Saponification value	25 – 50	25- 80
Saponification value of acetylated oil	125 – 155	145- 200
Colour of oil	Brown	Reddish green

Characteristic aroma of vetiver oil is due to presence of a number of ketonic sesquiterpenes especially vetiverol, α vetivone and β vetivone and also due to ester vetiveryl vetivenate. Sesquiterpene alcohols such as vetiverol and also vetivol acetate contribute to fixative property of oil. Oil contains more than 150 complex compounds including elemol 0.4-2.3%, α-epi-eudesmol 1.1-2.2%, β-eudesmol 5.5-8.5%, vetiverol+cyclocopacamphenol 6.1-7.5%; vetiselinenol 11-20%; khusimol 13-28%; β-vetivone 2-5%, α-vetivone 1.5-5.8%.

References

Adams, D. R. and Bhatnagar, S. P.1975. Analysis of the volatile constituents of essential oil of Indian linaloe. Ind.Flavours Food Addit., 6: 185-188

Aggarwal, K. K., Tickkoo, C. K., Narasimha, B. M. and Atal, C. K., 1973. Optimization studies on the production of cedar wood oil from stump and roots of Himalayan Deodar. Res. Ind., 18: 135-37

Ahmad, A. 1986. Isolation of new terpenoids from essential oils of potential value and elucidation of their structures. Ph.D. Thesis, Avadh University, Faizabad

Apte, U.B. and Jadav, B.B. 2002. Influence of different light intensities and fertilizers on growth and yield of Patchouli (*Pogostemon cablin* Benth.). FAFAI, 4: 51-53

Atal, C.K and Kapur, B.M. 1982. *Cultivation and Utilization of Aromatic Plants*. Regional Research Laboratory, Jammu-Tawi, 815p

Balz, R. 1999. *The Healing Power of Essential Oils*, Motilal Banarsidass Publishers Pvt. Ltd., Delhi, 203p

Bandyopadhyay, C. 1989. On some chemical aspects of *Cinnamomum zeylanicum*. PAFAI J., 11:35-38

Baruah, A.K.S., Bhagat, S.D., Hazarika, J.N. and Sasikia, B.K. 1975. Examination of volatile oil of *Cinnamomum camphora* grown in Jorhat, Assam. Indian J. Pharm., 37: 39-41

Bavappa, K.V.A. and Ruettimann, R.A. 1981. *Cinnamon Cultivation and Processing*. Tech. Bull.5. UNDP/FAO Research Project on Minor Export crops. Dept. of Minor Export Crops, Sri Lanka

Bhasker, S., Kumar, T. V. and Srivastava, H. C. 1997. Influence of growth regulators on production of herbage and oil in Patchouli (*Pogostemon patchouli*). Indian Perfumer, 41: 98-101

Bhasker, S.T and Kumar, V. 2000. Agronomic, bottlenecks, genetic barriers, and marketing impediments in Patchouli (*Pogostemon patchouli*) production. J. Med. and Arom. Pl. Sci., 22:396-403

Bisarya, S.C. and Dev, S. 1964. Allohima chalol, a new type in sesquiterpenoids. Tetrahedron Letters, 49:3761-3765

Bisarya, S.C., Shankaranarayanan, R., Krishnappa, S. and Dev, S. 1973. Deodarone, anevel sesquitarpene tetrahydro-pyrone from *Cedrus deodara* essential oil. Tetrahedron Letters, 6: 427-28

BOSTID, 1993. *Vetiver Grass: A Thin Green Line Against Erosion*. Board on Science and Technology for International Development. National Academy Press, Washington, D.C., 171p

Brown, E. and Matthews, W.S.A. 1951. Notes on the aromatic grasses of commercial importance. Col. Pl. Ani. Prod., 2: 174-187

Chandra, V., Misra, P. N. and Singh, A. 1979. *Matricaria for blue oil*. Extension bulletin No. 4. Economic Botany Information Service, National Botanical Research Institute, Lucknow, India

Chandra, V., Misra, P.N. and Kapoor, L.D. 1970. Blue oil from *Matricaria chamomilla* L.: A new industrial crop. Indian Perfumer, 14:27-31

Ciaramello, D., Azini, A., Pinto, A.J.D., Guilhero, M. and Danalisio, R. 1972. Use of citronella, lemongrass, palmarosa and vetiver leaves for the production of cellulose and paper. Anais Acad. Bras. Cienc., 44: 430-441

CIMAP, 2004. *Citronella Cultivation*. Farm bulletin, Central Institute of Medicinal and Aromatic Plants, Lucknow, 26 p

CIMAP, 2004. *Pogostemon patchouli Cultivation*. Farm bulletin, Central Institute of Medicinal and Aromatic Plants, Lucknow, 22p

Dhingra, S.N., Dhingra, D.R. and Gupta, G.N. 1955. Essential oil of *Ocimum basilicum*. Indian Oil Soap J., 19: 251-258

Doraswamy, K. 1967. Curing and distillation of patchouli. Indian Oil Soap J., 32: 211-216

Dutta, P.K. and Chatterjee, B.K 1961. Effect of harvesting intervals and nitrogen levels on the yield of *Mentha arvensis*. Sci. Cult., 27: 492-494

Dutta, P.K. and Singh, A. 1964. Effect of different spacings on fresh flower and oil yield of *Matricaria chamomilla*. Indian. J. Agron., 9: 11-12

Farooqi, A.A, Khan, M.M. and Vasundhara, M. 1999. *Production Technology of Medicinal and Aromatic Crops*. Natural Remedies Pvt. Ltd., Bangalore,176 p

Farooqi, A.A. and Sreeramu, B.S. 2001. *Cultivation of Medicinal and Aromatic Crops*, Hyderabad, Universities Press (India) Limited, 518p

Fuehrer, H. 1970. Vetiver oil and vetiver fragrances in modern perfumery. Dragoco Rep., 17: 217-219

Gangrade, S.K., Sharma R.K. and Pandya, S.C. 2002. Variability and correlation studies of different physico-chemical properties of essential oil of *Vetiveria zizanoides*. Indian Perfumer, 46: 165-168

GOI, 1986. *Report on the Marketing of Sandalwood in India*. Directorate of Marketing and Inspection, Ministry of Rural Development, Government of India. Faridabad, 196p.

GOI, 1988. *Production and Marketing of Palmarosa Grass and Oil in India*. Ministry of Agri., Dept. Rural Development, Directorate of Marketing and Inspection, Government of India, Faridabad,198p.

Grimshaw, R.G. and Helfer, L. 1995. *Vetiver Grass for Soil and Water Conservation, Land Rehabilitation and Embankment Stabilization*. Technical paper No. 273, World Bank, Washington, D.C., 281p

Grimshaw, R.G.1987. *Vetiver Grass* (*Vetiveria zizanioides*). Agriculture Division, World Bank, New Delhi. 72p

Guenther, E. 1950. *The Essential Oils*. Vol. IV. Van Nostrand Co. Inc., New York, 752p

Gulati, B.C. 1980. Essential oil of *Artemisia pallens* Wall. (Davana). Indian Perfumer, 24: 101-109

Gupta, R. and Singh, B.P. 2002. Crop maturity based on growing degree days in *Matricaria chamomilla* under residual sodium carbonate and zinc treatments. FAFAI J., 4: 41-44

Handa, K.L., Smith, D.M., Nigam, I.C. and Levi, L. 1964. Essential oils and their constituents XXIII. Chemotaxonomy of genus Mentha. J. Pharm. Sci., 53:1407-1409

Handa, S.S. and Kaul, M.K. 2001. *Supplement to Cultivation and Utilization of Aromatic Plants*, Regional Research Laboratory, Jammu-Tawi, 352p

Hiran, N. and Izawa, I. 1980. The constituents of the essential oil of Himalaya-Sugi (*Cedrus deodara* Loud) planted in Higa. Shiosaka-shikinki Daigaka Pikgakubu kenkyu kenky*u*, 15:61-64

Husain, A. and Ali, A. 1995. Endomycorrhizal (vesicular-arbuscular) infection in lemongrass *C. citratus* (DC). Stapf. Hamdard Medicus, 38: 63-70

Husain, A., Virmani, O.P., Sharma, A., Kumar, A. and Misra, L.N. 1988. *Major Essential Oil Bearing Plants of India*. Central Institute of Medicinal and Aromatic Plants, Lucknow, India, 237p

ISI, 1965. IS: 3398-*Oil of Patchouli*. Indian Standard Institute, New Delhi, India

ISI, 1969. IS: 1177-*Oil of Vetiver Roots*. Indian Standard Institute, New Delhi, India

Jadav, S.G., Jadav, B.B. and Apte, U.B. 2002. Effect of different growth regulators on vegetative propagation of Patchouli (*Pogostemon cablin* Benth.). FAFAI, 4: 25-27

Jadhav, S.G., Jadhav, B.B. and Apte, U.B. 2003. Influence of growth regulators on growth and oil content of Patchouli (*Pogostemon cablin* Benth). Indian Perfumer, 47: 287-289

Joy, P.P, Thomas, J., Mathew, S. and Ibrahim, K.K. 1998. Growth, leaf oil yield and quality investigations in cinnamon (*Cinnamomum verum*). J. Med. Arom. Pl. Sci., 20: 401-406

Joy, P.P., Skaria, B.P., Mathew, S. and Mathew, G. 2005. Cinnamon for flavour and aroma. Ind. J. Arecanut, Spices and Med. Plants, 7:1-6

Joy, P.P., Thomas, J., Mathew, S., Jose, G. and Joseph, J. 2001. Aromatic plants. In: *Tropical Horticulture* Vol. 2. Bose T.K., Kabir J., Das P. and Joy P.P.(eds.). Naya Prokash, Calcutta, India, pp 633-733

Junimann, M. 1999. *Enchanting Scents*. Sri Satguru Publication, Delhi, 123p

Kapoor, L.D. and Chandra, V. 1969. *Matricaria chamomilla* L.: A new essential oil yielding crop for India. Indian Drugs, 6: 55-61

KAU, 2003. *Package of Practices Recommendations 'Crops'-2003*. Directorate of Extension, Kerala Agricultural University, Thrissur, 278 p

Krishna Rao, G.S., Dev, S. and Guha, P. C. 1952. Studies in sesquiterpenes-Sesquiterpenes of essential oil from the wood of Himalaya deodar. J. Indian Chem. Soc., 29: 721-30

Krishnamoorthy, B., Rema, J., Zacharia, T.J., Abraham, J. and Gopalam, A. 1996. Navashree and Nithyashree - two new high yielding and high quality cinnamon (*Cinnamomum verum* Bercht and Presl.) selections. J. Spices and Aromatic Crops, 5: 28-33

Krishnan, R. 1981. Natural out crossing in sweet basil- *Ocimum basilicum* L. Indian Perfumer, 25 :74-77

Kumar, A., Gauniyal, A.K. and Virmani, O.P. 1986. Cultivation of *Pogostemon patchouli* for its oil. Curr. Res. Med. Arom. Plants, 8: 79-86

Kumar, V.K. 2004. Patchouli and India – A great leap forward. FAFAI, 6: 59

Lal, R.K., Sharma, J.R. and Sharma, S. 2000. Variability and stability pattern for some economic traits in Chamomile *(Chamomila reticulate)*. J. Med. and Arom. Plant Sci., 22: 219-222

Maheswari, P. and Biswas, C. 1970. *Cedrus.* Botanical Monograph No. 5, Council of Scientific and Industrial Research, New Delhi, India

Mukherjee, B.D., Light, K.K. and Hill, I.D. 1981. *A Study on Odour-Structure Relationship of Patchouli Compounds in Essential Oils.* Allured Publishing Corporation, Wheaton.

Nair, K.K.N. and Jayakumar, R. 1999. Ethno botany of Hill- Pulaya tribe in the context of biodiversity rehabilitation at Chinnar wildlife sanctuary, Western Ghats of India. J. Econ. and Taxonomic Bot., 23: 431-449

Narayana, K. and Dutt, S. 1945. Mysore linaloe leaf oil. Indian Oil Soap J., 10: 47-50

Narayana, M.R., Khan, M.N.A. and Dimiri, B.P. 1978. *Davana and its Cultivation in India.* Farm Bulletin No.12. Central Institute of Medicinal and Aromatic Plants, Lucknow, India

Nigam, M.C. Siddiqi, M.S. and Raychoudhuri, S.P. 1983. Status report on sandalwood oil. Curr. Res. Med. Arom. Plants, 5:124-138.

Pareek, S.K. and Gupta, R. 1985. On the status of agronomic research in *Cymbopogon* grasses in India with projections on future work. Indian Perfumer, 29: 215-224

Patra, D.D., Chand, S., Singh, A., Anwar, M., Lal, R. K., Singh, S., Bahl, J. R. and Khanuja, S. P. S. 2004. Agrotechnology of vetiver [(*Vetiveria zizanoides* (L) Nash]. J. Med. Arom. Plant Sci., 26: 784- 789

Patra, N.K. and Sharma, J.R. 1989. Genetic improvement of aromatic grasses: A review. Curr. Res. Med. Arom. Plants, 11: 32-39

Patra, N.K., Kumar, S., Khanuja, S.P.S., Shasany, A.K., Singh, H.P., Singh, V.R., Tanveer, H., Kalra, A., Singh, H.B. and Mengi, N. 1999. Genetic improvement of the cultivated species of *Cymbopogon* and *Mentha* for essential oil yield, and quality and adaptation. J. Med. and Arom. Plant Sci., 21:28-29

Prasad, A. and Singh, D.V. 2000. Effect of sodium – adsorption ratio of irrigation water on soil sodication and yield and on cation accumulation of Vetiver (*Vetiveria zizanoides*). Indian Perfumer, 44: 291-298

Radhakrishnan, V.V. 1992. Cinnamon - the spicy bark. Spice India, 5: 11-13

Rajendra, B. and Gupta, G.N. 1964. Design of a steam distillation equipment for vetiver (khas) oil. Indian Perfumer, 8: 87-90

Ram, B. and Misra, P.N. 2004. Nutrient accumulation, sodicity and reclamation potential of German chamomile *(Chamomila reticulata)* under varying sodicity and fertility levels. J. Med. and Arom. Plant Sci., 26 : 12-16

Ram, M., Ram, D., Singh, S., Nagavi, A.A. and Kumar, S. 1999. Studies on intercropping of Patchouli (*Pogostemon Patchouli*) with papaya (*Carica papaya*) . J. Med. and Arom. Plant Sci, 21: 358-360

Ramachandra, K.M., Vasundhara, M., Farooqi, A.A. and Srinivasappa, K.N. 2002. Evaluation of patchouli (*Pogostemon cablin* pellet) varieties in relation to different plant densities. Indian Perfumer, 46: 7-14

Ramachandra, K.M., Vasundhara, M., Farooqi, A.A. and Srinivasappa, K.M. 2003. Effect of varieties and spacing on growth, yield and quality of Patchouli (*Pogostemon Patchouli* Pellets). J. Spices and Aromatic Crops,12: 43-46

Ranade, G. 2002. Aromatherapy with special reference to lavender. Indian Perfumer 46: 272-275

Rao, I.K.S. 1964. *The Vetiver in India.* Indian Council of Agricultural Research, New Delhi, 40p

Rao, P.E.V.S., Rao, G.R.S., Narayana, M.R. and Ramesh, S. 1997. Influence of shade on yield and quality of patchouli (*Pogostemon cablin*) oil. Indian Perfumer, 41: 164-166

Sabeb, R., Menon, A.K. and Ittyachan, C.T. 1945. *Survey of Indian Vetiver (Khus) and its Oil.* CSIR Monograph. Council of Scientific and Industrial Research, New Delhi, India, 47 p

Sastry, K.P. and Kumar, S. 2000. Cultivation of economically important Eucalyptus species in the western ghats of Tamil Nadu. J. Med. and Arom. Plant Sci., 22: 528-531

Sastry, S.G. 1952. *Indian (Mysore) Linaloe Oil- A monograph.* Board of Scientific and Industrial Research, Govt. of Mysore, pp.1-12

Shamasundaran, K.S. and Bhaskar, S. 1997. Optimum size and shape plots for Patchouli (*Pogostemon patchouli*). Indian Perfumer, 41: 140-142

Shankaranarayan, V.G.M. 2002. Patchouli constituents and its usage in perfumery. Indian Perfumer, 46: 313-314

Sharma, A. and Sharma, T.C. 2003. Patchouli oil recovery and effect of leaf ageing. Indian Perfumer, 47: 151-154

Sharma, A., Kumar, A. and Virmani, O.P. 1983. Cultivation of German chamomile: A review. Curr. Res. Med. Arom. Plants, 5: 269-278

Sharma, A., Tewari, R. and Virmani, O.P. 1987. French Basil (*Ocimum basilicum* l.) A review. Curr. Res. Med. Arom. Plants, 9:136-151

Sharma, N. 1999. Conservation of Patchouli (*Pogostemon Patchouli*) through in vitro methods. Indian Perfumer, 43: 19-22

Shawl, A.S., Kumar, T., Shabir, S., Chishti, N. and Kaloo, Z.A. 2005. Lavender - a versatile industrial crop in Kashmir. Indian Perfumer 49: 235-238.

Shellie, R., Mandello, L., Morhott, P. and Dugo, G. 2002. Characterization of Lavender oils using gas chromatography mass spectrometry with correlation of linear retention and comparison with comprehensive 2 D gas chromatography. J. Chromatography, 970: 225-234.

Shinozaki, E. 1919. Japanese peppermint. J. Chem. Ind., 22: 296-312

Shrivasthava, D.K., Mishra, P.K. and Gangrade S.K. 2003. *Vetiveria zizanoides* Nash- A study of stability parameters. Indian Perfumer, 47: 165-168

Singh T., Handa, K.L. and Rao, P.R. 1976. Improved methods of maximum production of cedar wood oils. Parfum. Kosmet., 57:68-71

Singh T., Prabhakar, V.S. and Handa, K.L. 1964. Optimum conditions for production of Himalayan cedar wood oil. Res. Ind., 9: 291-95

Singh, D., Rao, S.M. and Tripathi, A.K. 1984. Cedar wood oil as a potent insecticidal oil as an agent against mosquitoes. Naturwissenschaften, 71: 265-66

Singh, M. 1999. Effect of irrigation and nitrogen levels on herbage and oil yield of patchouli (*Pogostemon patchouli*) on Alfsols. J. Med. and Arom. Plant Sci., 1: 689-691

Singh, M.1996. Effect of irrigation and plant spacing on herb and oil yields of Patchouli (*Pogostemon patchouli*) under semi arid tropical conditions. J. Med. and Arom. Plant Sci., 18: 487-488

Swamy, AY., Sreshtta, N.J. and Kalyanasundaram, S. 1960. Cultivation of scented geranium (*Pelargonium graveolens)* on the Nilgiris. Indian Perfumer, 4: 3-9

Thomas, J. 1995. Lemongrass. In: *Advances in Horticulture Vol. II- Medicinal and Aromatic Plants*. Chadha, K.L. and Guptha, R. (eds.). Malhotra Publishing House, New Delhi, 726p

Thomas, J. and Duethi, P.P. 2001. Cinnamon. In: *Handbook of Herbs and Spices*. Peter, K.V. (ed.). Woodhead Publishing Limited, Cambridge, England, pp.143-153

Thomas, J., Joy, P.P., Mathew, S., Skaria, B.P., Duethi, P.P. and Joseph, T.S. 2000. *Agronomic Practices for Aromatic and Medicinal plants.* Directorate of Arecanut and Spices Development, Calicut, 124p

TPI, 1961. *Vetiver Oil: Cultivation and Distribution*. Tropical Products Institute, London

Vadivel, E., Ponnuswami, V., Irulappan, I. and Dharmaraj, G. 1981. Vegetative propagation in cinnamon (*Cinnamomum zeylanicum*). South Indian Hort., 29: 231-232

Vaze, S. 1999. Essential oils - An independent retrospect and prospect. FAFAI J., 1: 17-19

Vijayalalitha, S.J. and Lade, R.R. 1996. A stimulation capacity of genotypes of patchouli (*Pogostemon patchouli* Pellet) in open and shade. Indian Perfumer, 40: 113-114

Weiss, E.A. 1997. *Essential Oil Crops.* CAB International, New York, 600p

Camphor (Page - 63)

Cedar Wood (Page - 66)

Chamomile (Page - 68)

Cinnamon (Page - 71)

Citronella (Java) (Page - 76)

Davana (Page - 82)

Eucalyptus (Page - 84)

Geranium (Page - 88)

Lavender (Page - 92)

Lemongrass (Page - 95)

Linaloe (Page - 105)

Mints (Page - 109)

Ocimums (Page - 114)

Palmarosa (Page - 118)

Patchouli (Page - 123)

Sandalwood (Page - 128)

Vetiver (Page - 131)

7

Other Sources of Aromatic Oils

Ajowan

Scientific name: *Trachyspermum ammi* (L.) Spragu; Syn: *T. roxburghianum* (DC.)

Family: Umbelliferae (Apiaceae)

Vernacular names: *Sanskrit*: Ajamoda; *Hindi*: Ajmud; *Malayalam*: Ayamodakam; *Tamil*: Asamatavomam; *Kannada*: Ajamodhavoma; *Telugu*: Ajumoda; *Gujarathi*: Ajmod

Ajowan is a profusely branched winter annual herb, seed oil of which is a major source of thymol, being present to extent of 35-60%. Ajowan oil is aromatic, stimulant and carminative. It possesses antimicrobial activity. Ajowan seeds are employed either alone or in combination with spices and condiments. It is used in pickles, confectionery and beverages. It is a good remedy for indigestion. A paste of crushed fruit is applied externally for relieving colic pains. It is also used in lotions and ointments.

Habitat and distribution

Plant originated in Egypt. It is cultivated around Mediterranean sea and in south-west Asia extending from Iraq to India. It grows wild in North India especially in Madhya Pradesh, Gujarat, Maharashtra, Uttar Pradesh, Rajasthan, Bihar and West Bengal.

The plant

Ajowan is a profusely branched winter, annual herb reaches up to a height of 60-90 cm. Stem is profusely branched and striated. Leaves pinnately divided and it has 7 pairs of lateral leaflets. Inflorescence is a compound umbel with 16 umbellets, each containing up to 16 flowers; flowers actinomorphic, white, male and bisexual; corolla 5; petals bilobbed; stamens 5, alternating with petals; ovary inferior; stigma knob-like; fruit aromatic, ovoid, cordate, cremocarp with a persistent stylopodium.

Cultivation

It is mainly grown as a winter crop in sub-tropical and temperate climate. It grows on any soil type but performs the best in humus rich loamy soil. It is grown as a rainfed crop in heavy soils whereas it requires irrigation in light textured soils. It is generally propagated by seeds. Field is ploughed repeatedly during September-October, incorporating organic manures at 10-15 tonnes/ ha. Seeds are sown broadcast or drilled in rows 45 cm apart in November. Seed rate is 3-4 kg/ha. Irrigation is given immediately after sowing and later at 7-10 day's interval. Seeds germinate in 7-14 days. Broadcast crop may be thinned to a spacing of 30-45 cm. N, P_2O_5, K_2O and S are applied at 80, 30, 30 and 50 kg/ha, respectively for obtaining best yield. Weeding is generally done twice. Collar rot caused by *Sclerotium rolfsii* is observed in some pockets. Flowering starts in two 2 months time. Harvesting is done in February-March when flower heads turn brown. Harvested crop is dried, threshed and winnowed to separate clean seeds.

Extraction and utilisation

Dried seeds are crushed and distilled to obtain essential oil. Hydro or steam distillation is usually carried out. Seeds lose essential oil when stored for long time. On an average, dry seeds contain 2-4% oil. Pale yellowish-brown ajowan seed oil has a characteristic thyme odour with sharp burning taste and physico-chemical properties of the oil shows specific gravity at 15°C 0.910-0.930, refractive index at 20°C 1.498-1.504, optical rotation 0–5°, content of phenols 45-57% and solubility in 80% alcohol 1- 2.5 vol.

Characteristic odour of ajowan oil is due to high content of thymol. On standing, major portion of thymol gets crystallized. Other major constituents are α-pinene, *p*-cymene, dipentene, α-terpinene and carvacrol. Following monoterpenes constitute hydrocarbon portion of oil: α-pinene 1.8%, camphene 0.5%, β-pinene 3.5%, myrcene 0.3%, δ-3-carene 0.5%, limonene 5.1%, α-terpinene 34.9%.

Ambrette

Scientific name: *Abelmoschus moschatus* Linn. Syn: *Hibiscus abelmoschus* (Linn.), *A. rugosus* Wall.
Family: Malvaceae
Vernacular names: *Sanskrit*: Lata kasturika; *Hindi*: Guj; *Bengali*: Mushkdana; *Malayalam*: Kasthurivenda; *Tamil*: Varttilai-kasturi; *Kannada*: Kasturi-bende; *Telugu*: Kasturi-benda; *Gujarathi*: Binda

Ambrette is a native of India and comes up well throughout tropical regions of the country. It possesses delicate musk like odour, valued for volatile oil present in seed coat. Oil extracted from seed coat is called ambrette oil. India exports ambrette seeds to Japan, Saudi Arabia, U.S.A., Germany, France, U.K, Baharain, Italy, Greece, and Sri Lanka. Essence of ambrettolide is also exported to Bangladesh, Germany, Nepal, South Africa, Switzerland, Sri Lanka, U.A.E and U.S.A. India imports musk ambreth from Germany, Israel, Japan, Malaysia, Netherlands, U.K and U.S.A. Being a foreign exchange earning crop, there is scope for cultivation of ambrette in India. Ambrette oil is extensively used in high class perfumes as 'masculine perfume', flavoring and cosmetic industries. Essential oil is present in seed coat. Seeds are used to impart a musky odour to sachets and hair powder and in manufacture of indigenous flavoured tobacco (Zarda). They are mixed with clove and other scented materials for use in baby perfumes. It is also used as a substitute for 'kasturi' or musk, an animal product. Ambrette restored in the form of extracts is used in perfumes creams, lipsticks, and brilliantine's hair oil and in cosmetic products. Seeds also possess medicinal properties and are employed in treatment of diseases due to 'Kapha' and in stomach and urinary diseases. They are stimulant, antiseptic, stomachic, cooling, tonic, carminative and aphrodisiac. Plant yields good quality fibre. Leaves are used for cleaning sugar. Tender leaves and shoots are used in soups and green tender pods are used as vegetable. Seeds also possess insecticidal properties. Its seed coat yields an aromatic "absolute" which can serve as a highly useful base for preparing high quality perfumes, scents and cosmetics. Seeds are also used as antispasmodic, tonic against hysteria and other nervous disorders, to protect woolen garments against moth and imparting musky odour to pan masala and agarbatties. They check vomiting and cure diseases due to 'Kapha' and 'vata'. Seeds are used as cardio tonic. Leaves and tender shoots are used to clean jaggery and tender pods used as vegetable.

Habitat and distribution

Ancestral homes of this species are India, south east Asia, South China, Indonesia, Peninsular Indo China, New Guinea, Northern Australia and the South West Pacific Island . The species is under cultivation in Seychelles, Madagascar, Columbia and Brazil. In India, it grows wild all over hilly regions of Deccan, Karnataka and in foot hills of Himalaya. Crop is suitable for cultivation as a cash crop in the states of Madhya Pradesh, Orissa and Maharashtra.

The plant

Ambrette is an exalbuminous, erect, hirsute, annual or biennial herb attaining a height up to 2 m. Stem is hollow and pubescent throughout. Leaves are polymorphous of varying shape usually palmate with 5 to 7 lobes. Flowers are solitary bright yellow coloured with a purple centre and large in size, 7.5 to 10 cm in diameter. Fruit is a dry, dehiscent capsule, oval to fusiform, slightly angled, usually five chambered, 5 to 7.5 cm long containing a large number of reniform pubescent seeds which are musk scented. Oil secretion begins when seeds start developing brown colouration on seed coat. Essential oil secreting structures are found in testa.

Cultivation

Ambrette is a hardy plant and comes up well on a wide range of soils. Sandy loam soil with a pH of 7 is the best for cultivation of ambrette. It also yields well in soil with a pH of 6 to 8.6. It prefers loose, fertile and well drained soils. Crop is cultivated widely in India up to an elevation of 1000 m under different climatic conditions. It occurs throughout hotter part of India. In Karnataka it can be grown twice a year, once during June–July and again in October-November as an irrigated crop. In Terai area of Kumaon (UP) and in Punjab, crop is grown during rainy season. Plant is propagated through seeds. Land should be well manured as the plant is a heavy feeder. Seeds can be sown twice during the year, during June-July and again in September-October. Seeds take about 8-10 days for germination. While sowing, 2-4 seeds may be sown on each hill by dibbling to a depth of 1 cm. Pre soaking seeds for 24 hours in water improves germination. About 6 kg of seeds are sufficient for sowing one hectare area. A spacing of 60 cm between rows and 30 cm between plants is recommended. Seedlings are thinned 20 days after sowing.

Well decomposed FYM is applied to soil @ 15 tonnes/ha. Since ambrette is a heavy feeder, chemical fertilizers are also applied in large quantities. Fertilizer trials showed that application of N at 120 kg, P_2O_5 at 35 kg and K_2O at 40 kg/ha resulted in best yields under Bangalore conditions. Of these, full dose of P and K and 40 kg N are applied as basal dose, while remaining 80 kg

N is applied in two equal split doses of 40 kg each at 60 days and 120 days after sowing. Fertilizer mixture is applied about 10 cm away from plants and mixed well into soil. Aphids, thrips, pink boll worm, spotted boll worm and red spider mites are major pests causing considerable damage to crop. Foliar spray of dimethoate, malathion and phosphomidon reduce risk of these pests. Several fungi and virus diseases like mosaic disease, anthracnose and leaf spot disease are common in ambrette. Repeated spray of ridomil (0.1%) controls infection. Plants infected with Hibiscus Mosaic Virus (HMV) should be uprooted and destroyed.

Pruning of plants 50-60 days after sowing induce early pod formation and more seed yield. Pruning of plants at a height of 50cm encourage branching. Early pruning results arrest of growth and production of sympodial branches and causes reduction in seed yield. Crop starts flowering after about 2½ moths of sowing. Flowers set fruits in nearly 3-4 days. Fruits mature in two months. Harvesting of pod is started when they turn blackish and white stripes appear at angles at ridges of fruit. Harvesting is prolonged if flowering season is long. Fruit should be harvested carefully by cutting stalk as it possesses stiff hairs which cause itching. Harvesting must be undertaken regularly at intervals of 7-10 days, depending upon availability of mature fruits. It is a crop of 170-180 days duration and in all 20–25 pluckings are to be carried out. Pods have to be shade dried after harvest. Seeds are separated from pods after beating pods with sticks or by splitting dry pods by hand. Seeds have to be cleaned, shade dried and stored. A normal crop gives a seed yield of 0.9-1.0 tonnes/ ha.

Pest and disease management

Pests

Many pests attack ambrette when crop is grown commercially. Aphids and thrips are major ones in early stages, especially in summer crop. Foliar sprays of dimethoate, mercaptothion or phosphamidon reduce risk of aphids and thrips. Nymphs and adults of leaf hopper are prevalent throughout the cropping season and maximum damage is caused during July-September and November–January. Severe infestation results in retardation of growth of plants and premature defoliation. Caterpillars of leaf roller, roll leaves from edge to mid rib forming a cover around it and feed inside which results in defoliation of leaves. Spotted boll worm attacks succulent shoots, vegetative buds, flowers and pods during vegetative growth and fruiting stage. Infested shoots, above the point of infestation become brown, bend down and die. Application of insecticides with stomach and contact action is effective against boll worms. Pink boll worms bore into pods. Healthy seeds free from infestation should be sown as a precautionary measure. Red spider mites feed from under surface of leaves and are seen in large numbers inside a silken web. Infested plants become yellow, weak and shed leaves. Miticides and insecticides like dimethoate or phosphamidon prevent damage. Red cotton bugs feeding on tender portions and seeds can be controlled by any contact insecticides. Against root knot nematodes infesting roots of ambrette, carbofuran 3G @ 2-2.25 kg a.i/ha can be applied.

Diseases

Ambrette is prone to several diseases. Plants infected with Hibiscus Mosaic Virus (HMV) should be uprooted and destroyed. Anthracnose disease starts from seedling stage and hence seeds should be treated with a fungicide before sowing. Fusarium wilt results in death of plants unless managed by spraying 1% Bordeaux Mixture. Leaf spot and leaf blight disease are common in crop especially during high humid conditions. Repeated sprayings with fungicide, zineb at 0.5% at monthly interval control blight effectively.

Extraction and utilisation

Oil is extracted from seed by steam distillation followed by solvent extraction. Concentrate of solvent extraction is further extracted with alcohol to get absolute which is alcohol soluble volatile concentrate. Volatile oil also known as musk seed oil or ambrette seed oil is obtained through hydro-distillation of crushed seeds using Clevenger apparatus. Fatty oil of seeds contains phospholipids 2-cephalin, phosphatidyl serine and its plasmalogen and phosphatidyl choline plasmalogen. Absolute contains farnesol and ambrettolic acid lactone. β-sitosterol and its β-D-glucosides are isolated from leaves. Petals contain β-sitosterol, flavonoid, myricetin and its glucoside. Anthocyanins like cyanidin-3-sambubioside and cyanidin-3-glucoside are present in the flowers.

On an average, ambrette seeds yielded 0.3% essential oil on hydro-distillation. Gas chromatography analysis data of essential oil revealed presence of 12 constituents. Major constituents of oil identified are farnesyl acetate (75.82-78.67%) and ambrettolide (7.83-9.21%).

Caraway

Scientific name: *Carum carvi* L.
Family: Apiaceae
Vernacular names: *English*: Caraway; *Hindi*: Siyazira; *Bengali*: Jira or Zira; *Kannada*: Shime jeeriji; *Malayalam*: Sheema jeerakam; *Panjabi*: Zira siah; *Sanskrit*: Sushavi

Caraway is a member of aromatic plants characterized by carminative properties. Other plants of the group are anise, cumin, dill and fennel. It is grown for medicinal properties of fruits and seeds. Fruit is an aromatic carminative. Caraway water is used as carminative for children. Caraway water also finds use in non alcoholic beverages, ice creams, bakery and condiments. Fruits are used for flavouring cakes, bread and various items of confectionery. It also finds use in non-alcoholic beverages, ice creams, bakery and condiments.

Caraway oil, the aromatic oil distilled from seed is used in flavoring bread, cheese, meat, pickles and sauces. It is also used in mouth washes, tooth paste, chewing gum and candies. It has a special use as masking agent in bad tasting pharmaceutical preparations. The essential oil is also employed in soap, perfumes and various cosmetics. The oil rich in carvone is used as carminative and stomachic to alleviate nauseating effect of some medicines. It is also recommended against dyspepsia, lysteria and indigestion.

Habitat and distribution

Plant is cultivated in northern and central Europe. Holland is the chief producer followed by Germany, Finland, Russia, Australia, France, Spain and Morocco. In India, it is growing wild in north Himalayan regions. Caraway is cultivated as a summer crop in Himalayan foot hills of Kashmir, Kumaon and Garwal at altitudes ranging from 2500 to 4000 m. In plains, it is grown as a winter crop.

The plant

Plant is an erect, herbaceous and biennial herb with thick tuberous root stocks. It grows up to a height of 150 cm. Stem is slender to robust, divaricately branched, aromatic, striate and leafy. Leaves are pinnately compound and ultimate segments of leaves are lanceolate. Flowers are minute, white, terminal or none, rarely divided. Bracteoles are small, linear or none. Calyx is 5-toothed, small or none, petals 5, notched, often enlarged and irregular. Carpels are rounded and narrowed upwards. Fruits are brown, cremocarpous, 3-6 mm long, ovoid or oblong, glabrous and laterally flattened. Seeds are dorsally flattened, smooth or slightly grooved on inner face. Fruits which are popularly and incorrectly called seeds, are laterally compressed, somewhat horny, and translucent, slightly curved and marked with five distinct pale ridges. Roots are thick and tapering like a parsnip and are edible.

Cultivation

Caraway grows well on variety of soils but well drained, humus-rich, sandy loams are preferred. Crop grows as an annual in the plains and as a perennial in the hills. Plants are raised either from seeds or from mature bulbs. Seeds are sown directly in field either by broadcasting or dibbled 2-3 cm deep and 35-60 cm apart; 4-5 kg seeds are required for planting one hectare. Crop is to be irrigated immediately after sowing. Seeds germinate within one to two weeks. Farmyard manure at 20-25 tonnes/ha and fertilizers @ 30 kg N, 12 kg P_2O_5 and 20 kg K_2O/ha are recommended.

Crop grown in plains flowers during first year itself. But in hills, aerial parts of plant die in July and bulbs reproduce during March. Inflorescence is produced during fourth year only. Weeding is done regularly but hoeing is avoided as it disturbs bulbs.

As in case of other umbelliferrous plants, extra care is required in harvesting fruits to avoid loss of seeds by shattering. Crop is mown when the oldest fruits turned brown. Sheaves are then stacked in piles of 20-30 to allow completion of ripening and drying. It is advisable to mow the crop in early morning or late evening hours when plants are wet with dew, collecting sheaves on tarpaulins. Sunny days are chosen for harvesting because seed then contains more oil. After being cut, plants are stacked up and left in fields for a week or ten days until dried. Threshing can be done as soon as crop is dry. Yield varies from 750 to 2000 kg/ha with an average of 1250 kg/ha.

Extraction and utilisation

Oil is extracted from dry seeds by steam distillation. Crushing of seeds before distillation helps to increase oil yield and quality. Crushed seeds are spread evenly on perforated grids provided in still to assure complete penetration of live steam. Optimum period of distillation is about 6-8 hours. Approximate oil yield is 3-6%. Spent seed cake contains about 20-23% crude protein and 14-16% fat and is suitable as a cattle feed.

Physico-chemical properties of oil

Appearance	Colourless to pale yellow
Specific gravity	0.90 - 0.92
Refractive index @ 25°C	1.484 - 1.498
Optical rotation	70° to 81°
Solubility in 80% alcohol	10 volumes.
Dextro carvone content	45-65%

Chemical constituents of caraway oil are carvone 47.1%, myrcene 0.6%, *trans*-carvol 0.4%, cis-carveol 0.2%, *trans*-dihyhydrocarvone 0.2%, *neo*-dihydrocarvone+germacrine-D 0.1%, α-pinene 90.1%, linalool 0.1%, sabinene 0.1% and trace amounts of β-pinene, thujene, camphene, Δ-3 carene, α-phyllandrene, terpineolene, octanal, nonanal, cis-*p*-menth-2-en-1-ol, terpinene-4-ol, dihydrocarvinol, *iso*-dihydrocarvinol, *neo*-isodihydrocarvinol and *cis*-*p*-menth-2,8-dien-1-ol.

Celery

Scientific name: *Apium graveolens* L.
Family: Umbelliferae (Apiaceae)
Vernacular names: *Sanskrit*: Ajmoda; *Hindi*: Badi ajmud; *Bengali*: Chanu; *Malayalam*: Celery; *Punjabi*: Bhutghata

Celery popularly known as Karnauli *or Ajmod* is an annual or biennial erect herb with jointed stems whose seed on distillation gives a pale-yellow essential oil which is used as an essence in flavour and pharmaceutical industry. Bulk of demand comes from canned soup industry. It is used in flavouring of all kinds of prepared foods such as soups, meats, pickles, vegetable juices and in preservation of meat sauces. In pharmacy, oil is used in certain preparations having sedative effect. It is highly priced for fixative purposes and as an ingredient of novel perfumes. It has a powerful odour and imparts a pleasant warm note. Oil is used in compounding ayurvedic formulations. Fruits yield 17% fatty oil which is used as an antispasmodic and nerve stimulant. Seeds of celery are rich in vitamin B.

Habitat and distribution

Celery is a native of northern hemisphere extending from Sweden to Egypt. It is cultivated in Algeria, Ethiopia, India, Caucasus, Baluchistan, France, Holland, Hungary, China and U.S.A. In India, it is mainly grown in Punjab, Haryana and Uttar Pradesh.

The plant

Two varieties are recognized under the genus, *Apium.*

A. graveolens L. var. *dulce* (Mill.) Pers. Leaves and flowering stems are used as appetizer and salad at table and as a flavouring agent in soups.

A. graveolens L. var. *rapaceum* (Mill.) Gandich or *turnip-rooted celery.* This is smaller with dark green leaves, less developed stalks and swollen roots (5-6 cm in diameter) which are eaten after cooking.

Celery is an annual in plains producing seeds in the very first year but in colder climates and on hills it becomes biennial and produces seeds only in second year. It is an erect herb of 60-90 cm height. Roots are succulent and numerous. Stems are branched, angular or fistular, conspicuously jointed.

Leaves are pinnate, deeply divided into three leaflets which are ovate to suborbicular and 3-lobbed; inflorescence compound umbel; flowers small, white; calyx teeth obsolete; petals 5, ovate, acute with tip inflexed; carpels semiterete, filiform; fruits schizocarp with two mericarps, aromatic and slightly bitter.

Cultivation

Celery prefers a moist, cool climate. It grows as an annual in plains but as a biennial at higher elevations with cooler climate. It flourishes well on fertile, well drained, sandy and silt loam soils. Clayey soils are not suitable.

Plant is propagated through seeds. Seeds obtained from primary umbels are heavy and produce better seedlings in comparison to seeds obtained from quarternary umbels. On hills or higher elevations, seeds are sown during March-April, transplanted in May and harvested in November. In plains, seeds are sown during September-October, transplanted in January and harvested in May. High temperature pretreatment of seeds increases germination rate. Healthy seedlings are obtained by incubating seeds at 90% relative humidity and 15-20°C for 8-10 days. For transplanting one hectare, 1.5 kg seeds are sown in a nursery area of about 1000 m^2.

Main field is thoroughly prepared incorporating organic manures at 10-20 tonnes/ha according to availability and transplanting is done in moist soil at 30-40 cm spacing. Fertilizer application at 200 kg N and 40 kg P_2O_5/ha is much remunerative. Crop needs plenty of water and field is irrigated every 10-15 days during non-rainy period. Crop may lodge when strong winds blow after irrigation which may be prevented by providing wind breaks. Field is kept weed free by 2-3 hoeings, first 3 weeks after transplanting and subsequently at 2 weeks interval.

Crop is harvested when white flowers start turning reddish. Harvested crop is thrashed with sticks next day and seeds are taken. Average seed yield is 1-1.5 tonnes/ha.

Pest and disease management

A leaf miner, *Liriomyza trifolii* and a fungus *Septoria apicola* attack crop. They can be controlled by suitable insecticide and fungicide, respectively.

Extraction and utilisation

Celery seed oil is obtained by steam distillation of seed. Usually, distillation is carried out for 18 hours. Oil is very light and hence separation from water is not a problem. Celery seed contains 2-3% of essential oil. From chaff also an essential oil can be obtained, which of course, lacks better aroma of seed oil. Celery chaff oil and synthetic d-limonene are common adulterants of celery seed oil which are difficult to detect.

Physico-chemical properties of oil

Celery seed oil is a light pale-yellow coloured liquid with persistent characteristic odour of plant.

Specific gravity at 15°C	0.866-0.898
Refractive index at 20°C	1.478-1.486
Optical rotation	51° to 82°
Acid value	upto 4.0
Ester value	16.0-55.0
Ester value after acetylation	43.0-67.0
Solubility in 90% alcohol	6-8 vol.

Composition of celery seed oil was reported as limonene 80.0%, α-p-dimethyl styrene 0.9%, n-pentyl benzene 1.0%, caryophyllene 0.5%, α-selinene 0.5%, n-butyl phthalide 1.0% and sedanolide 0.5%. Other constituents are sabinene, β-elemene, trans-1,2-epoxy limone, linalool, iso-valeric acid, cis-and trans-dihydro carvone, terpinen-4-ol, cis-and-trans p-menth-2, 8-dien-1-ol, α-terpineol, carvone, cis-and-trans carveol, trans anethole, trans-carvyl acetate, cis-and-trans-p-menth-1(7), 8-dien-2-ol, perillaldehyde and thymol. Celery leaf oil is richer in mono and sesquiterpenes in comparison to celery seed oil.

Deterpenation of volatile oil gave about 13% terpeneless oil. Yield of celery seed oleoresin was about 24% and oleoresin kept well when stored in cold (8-10°C) for 60 days even without addition of antioxidants.

Champak

Scientific name: *Michelia champaca* Linn.
Family: Magnoliaceae
Vernacular names: *English*: Golden yellow champa; *Sanskrit*: Champak, Hemapushpa, Pitapushpa, Sthiragandha, Gandh phalli; *Hindi*: Champa, Champaka; *Bengali*: Chanpa; *Tamil*: Shampgi; *Malayalam*: Champakam; *Telugu*: Sampangi; *Gujarathi*: Raichampa, Peelo champo; *Kannada*: Sampge; *Marathi*: Sona champa, Pivala champa

Champak, Champa or Yellow Champa is a large evergreen tree valued for its beautiful flowers with long-lasting fragrance. Tree is a great favourite in Hindu gardens, exquisitely scented flowers being used for *pooja* particularly of Lord Krishna. Besides, ladies are very fond of champak flowers because of its pleasant fragrance. Champak attars are produced in India which are used in hair oils as a head coolant. Champak oil is highly esteemed in perfumery and traded in international market. Flowers also yield a dye, used as a base for other colours and for dyeing silk and cotton fabrics. By virtue of refreshing appearance of its foliage, champak looks elegant even when it is out of flowers. Leaves when distilled yield scented water. All parts of plant have medicinal properties and are useful in treatment of various ailments like inflammation, constipation, gastritis, fever, cough, bronchitis, cardiac debility, ophthalmic disorders and urinary problems. Wood takes good polish and is used for house and carriage building and for furniture.

Habitat and distribution

Champak tree originated in south east Asia. By way of habitation it is found distributed in India and Nepal. It is commonly cultivated, but mainly found wild in forests. In India, it is distributed over eastern Himalayas, North East Assam, Western Ghats, Nilgiris, Tamil Nadu, Southern Orissa, West Bengal and also in Nepal. Tree thrives the best in damp climate and requires deep moist soil. It is a moderate light demander and is sensitive to frost. It requires a mild climate and an elevation of 100-1000 m with partial shade for good growth. It can be grown on a wide variety of soils, however, well drained rich sandy loam soils are the best.

The plant

Plant is a medium sized tree with oblong crown that grows to a height of 30 m. Leaves are ovate to lanceolate of size 20x6.5 cm. Tree flowers in hot and rainy weather and seeds late in August. Buds are silky. Flowers yellow to orange, fragrant, solitary, axillary and bell shaped about 5-6.2 cm diameter, each enclosed in bud by a greyish yellow, pubescent, spathaceous, coriaceous bract, sepals and petals 15 or more, stout, wrinkled, marked with annular scar, round the middle. 'Simhachalam' golden orange type is the most sweet scented and is the most favoured of champaks. White flowered champak, though very sweet scented, lacks in substance and hence fragrance does not last quite long. Fruits are 5-10 cm long; ripe capsules ovoid or ellipsoid, valves woody. Seeds 1-12, brown, angular with pink fleshy aril. Natural regeneration is usually plentiful around mother trees.

Cultivation

Trees are propagated both by seeds and vegetatively by grafting. Creamy yellow variety is propagated by grafting on stocks of ordinary golden orange flowered variety which produces seeds in bunches and takes 7-8 years to flower. Seedlings of 'Simhachalam' golden orange type bear flowers in about 3 years and produce taller trees than grafts which bear fruits in a short period. Though large scale commercial cultivation of champak is not common, group planting is generally undertaken, particularly in informal gardens and homesteads. Artificial reproduction is accomplished by sowing seeds in nursery and transplanting 12-15 months old seedlings. Seedlings are pretreated with Gibberilic acid for better germination and sown at 1.5 cm depth in nursery beds. Germination commences after 38 days and completes within 70 days. Care is to be taken till grafts or seedlings are initially established in field after which not much attention is needed. Trees flower during April-May and again during September-October once they start blooming. A well grown tree yields 50-100 flowers daily during peak season; 375 to 425 flowers weigh one kilogram.

Extraction and utilisation

Champak flowers are exquisitely fragrant. Owing to presence of an oxidizing agent in flowers they become brown within a few hours after picking and are subject to odour deterioration. To prevent impairment of its odour by oxidation, essential oil must be extracted soon after picking. The volatile oil is not generally extracted by steam distillation because of poor yield and odour of that oil having no resemblance to that of the flowers. Concrete yield by solvent extraction is around 0.26% which in turn is capable of yielding 26% of steam volatile oil. Enfleuraged flowers in sesame oil yield excellent attar.

Champak flower concrete is a waxy solid having a low melting point of 29°C. Champak absolute has a very sweet smooth floral and velvety odour, closely resembling that of the live flowers. Champak oil has low saponification value and low solubility in 90% alcohol. The oil does not resinify during fractionation. It contains important perfumery constituents such as cineole, iso-eugenol, phenyl ethyl alcohol, benzaldehyde, methyl anthranilate, benzyl alcohol.

Clarysage

Scientific name: *Salvia sclarea* L.
Family: Lamiaceae (Labiatae)

Clarysage is a perennial herbaceous plant. Flowering tops, leaves and derivatives of this plant are extensively used in flavour industry for formulation of liquors and soft beverages. Plant was used in middle ages for clearing vision and for this reason it received its popular name, clarysage meaning clary-eye. Clary is chiefly known for its aromatic oil which is used in manufacture of perfumes, soaps and cosmetics. Essential oil which bears coriander-like notes is used in perfumery. It is used as a flavour in liquors and as a modifier of spice compound preparations. Oil is also used in preparation of non-alcoholic and alcoholic beverages, ice-creams, candy and baked goods.

Habitat and distribution

Clarysage is a native of Mediterranean region and middle east. It is grown extensively in Russia, U.S.A., Bulgaria, Yugoslavia, France, Switzerland and Morocco. In India, all areas in Kashmir valley especially the uplands and dry dunes in hilly tracts of Himachal Pradesh, Uttar Pradesh and Uttaranchal are suitable for its cultivation. Major consumers are U.S.A, U.K., Italy, Canada, France and Japan.

The plant

Sage oils are obtained from different species of *Salvia*. Inferior Spanish and Dalmatian materials are obtained from *S. officinalis* L., whereas superior and more expensive oil is obtained from *S. sclarea* L. Latter is the commonly cultivated type. *Salvia* is a 60-90 cm tall herb with a deep and widely branched root system. Stem is branched, 4-ribbed with numerous epidermal glands on younger parts; leaves petiolate, opposite, broadly oval, toothed and thinly hairy above; flowers many, blue to purple, in 15-80 cm long peduncle, bisexual and zygomorphic; calyx gamosepalous, inferior; corolla bilipped, upper 3-lobed, lower 2-lobed; stamens two, inserted in corolla tube; ovary bicarpellary, syncarpous; superior; stigma bilobed; placentation axile. Honey discs are present at base of ovary and 138 kg/ha of honey has been reported to be recovered from these honey discs. Anisophylly is observed in clarysage. *Xylocopa violacea* is reported as the only effective pollinator of clarysage in Yugoslavia. Cells of pericarp and skin of fruit contain oil drops. Fruits differ in size, weight, colour, swelling intensity and mucilization.

Cultivation

Clarysage is tolerant to cold and drought and adaptable to a wide variety of situations. Higher altitude with ample sunshine and a few good showers in spring results in good yield of oil having superior quality. Plant is generally grown on poor soils. Slightly acidic soils of pH 4.0-5.5 are the best suited. Crop is propagated through seeds. A few high yielding hybrid varieties have been developed in Bulgaria . 'Zarya' is a medium early variety with 0.24% oil and 78% linanyl acetate in oil. 'Lazur" is cold resistant with 0.23% oil and 73% linalyl acetate. Seeds can be directly sown in field or transplanted either in November or March-April depending upon weather conditions. Seed rate is 3-4 kg/ha for transplanting. Seedlings appear in 10-15 days and are transplanted when 30-35 days old, at 1 m row spacing after incorporating 10-12 tonnes of organic manure in field for optimum growth of plants. 100-120 kg N and 30 kg each of P_2O_5 and K_2O are recommended/hectare. N may be applied in 4 equal splits. One or two irrigations may be given in drought situation. One or two weedings should be done during March-April. Pre-emergence application of fluometuron or diuron at 2 kg/ha and post-emergence application of preforan or introchlor at 3 kg/ha are recommended for weed control. Two or three hoeings should be done before flowering season.

Flowering tops and leaves are harvested twice a year during July and September. Excessive stalk growth is removed as it contains no significant amount of oil. After harvest, a hoeing is given. Plants remain productive for 5-6 years after which there is considerable yield decline. Thereafter new plantation is started in a different location.

Pest and disease management

An aphid, *Acyrtosiphon salviae* is found in colonies on clarysage which can be controlled by a mild insecticide. Rootknot nematode *Meloidogyne incognita* infests the plant heavily.

Fungus *Rhizoctonia solani* causes root rot disease. Under humid and wet conditions, whole plant collapses within 2-3 days. Drenching with copper oxychloride or 1% Bordeaux Mixture is recommended.

Extraction and utilisation

Harvested herb is distilled immediately to avoid evaporation loss of essential oil. Distillation is carried out for a period 2-3 hours using live steam from a boiler. Recovery of oil from herb grown on poor soils is about 0.15%. However higher recoveries of 0.2-0.3% are obtained with improved varieties and good management in which case, oil yield will be around 40-50 kg/ha.

Physico-chemical properties of oil

Oil of clarysage is an yellow liquid with a characteristic herbaceous odour, wine like taste and following properties.

Specific gravity (20°C)	0.89-0.90
Refractive index (20°C)	1.4675-1.4710
Optical rotation (21°C)	-15° 12' to -29° 3'
Acid value	0.61-1.20
Ester content calculated as linalyl acetate	26.50-45.34%
Saponification value	105.18-129.60
Total alcohol content calculated as linalool	60.20-68.01%
Solubility in 85% alcohol	0.6-1.20 vol.

Chemical constituents of oil are reported to be α and β-ocimene, *p*-cymene, terpinolene, *cis*-3-hexen-1-ol, linalool (the major constituent) and its acetate, terpin-4-ol, caryophyllene, α-terpineol, citronellol, nerol, geraniol, and their acetates, *trans*-β-terpineol, β-gurjunene, caryophyllene oxide, tricyclene, camphene, 1,8-cineole, methyl heptanone, camphor, β-thujone, β-humulene, α-thujone, δ-cadinene and citral a and b.

Dill

Scientific name: *Anethum graveolens* L.; Syn: *Peucedanum graveolens*

Family: Umbelliferae (Apiaceae)

Vernacular names: *Sanskrit*: Satapushpi; *Hindi*: Sowa, Soya; *Tamil*: Satakuppi; *Kannada*: Sabasige

Whole plant is aromatic, but most of volatile oil is present in seed. Dill herb oil and dill seed oil are obtained from herb and mature seeds. Former is obtained by steam distillation of herb including immature fruits whereas latter is obtained from mature separated fruits. Leaves are used as a culinary herb for seasoning soups, sauces and particularly pickles. Seed is used as a condiment in India. Dill seed is excellent in pickles, sauerkraut, sauces, salads, soup, pie, stews and cheeses. Oil is of very high demand in food, pharmaceutical, perfume, flavour, cosmetic and soap industries. It is used in baked goods, ice creams, chewing gums and beverages. Rectified oil can replace caraway in perfumery.

Dill is carminative and diuretic in ayurvedic and unani medicines. Dill oil relieves intestinal spasms, relieves pain and induces sleep. It is used in cough, cold and flu remedies and is a mild diuretic and galactogogue. Emulsion of seed oil is useful in flatulence, colic pain, vomiting and hiccups. European dill seed oil is preferred in Pharmacy over that of Indian dill due to presence of Dillapiole ranging from 15.6 to 39.6% and lower levels of carvone. Dill seed oil possessed potent antibacterial activity against human pathogenic bacteria viz. *Escherichia coli, Salmonella typhi, Proteus vulgaris, Staphylococcus aureus* and *Pseudomonas aeruginosa*.

Habitat and distribution

It is a native of southern Europe and central and southern Asia. It is cultivated in England, Denmark, Hungary, Austria, Germany, Holland, Bulgaria, France, Greece, Poland, Sweden, Russia, India, Canada, Mexico, Pakistan, Romania and the United States. In India, its cultivation is mainly confined to Jammu and Kashmir, Maharashtra, Gujarat, Andhra Pradesh, Madhya Pradesh and Rajastan. Indian Dill (*Anethum sowa* DC) is found throughout tropical and subtropical India and often cultivated for use as a vegetable and also as a source of essential oil.

The plant

European dill (*Anethum graveolens* L.) is an annual or occasionally biennial herb growing to about 90 cm in height (2n= 22). It has a 10-16 cm long fusiform tap root, an erect, hollow, smooth dichotomously branched, cylindrical stem and feathery, linear, up to 50 cm long and 28 cm broad tripinnate leaves with pointed leaflets and a clasping base. Plant has a bushy appearance due to many axillary branches arising from lower nodes. Flowers are numerous, yellow, regular, bisexual, pentamerous, borne on compound umbels terminally. Petals are small, yellow and rolled inwards. Fruit is a cremocarp, very pungent and bitter, flat, oblong, dry, brown in colour and light in weight but produced in large numbers. Fruits are commonly, but erroneously called seeds.

Indian dill (*A. sowa* DC.) is a closely related species, native to northern India and is cultivated in Japan and South East Asia. It is cultivated on a large scale in India as a winter crop. Its oil contains lesser percentage of carvone than prescribed in Pharmacopoeia. It can be rectified and used as a substitute for European dill. In this species, ultimate segments of leaves are shorter. Fruit is less compressed dorsally and longitudinal ridges are more prominent.

Cultivation

Well drained, fertile, sandy loam soils are ideal for the crop though it comes up well in any garden soil. Saline soils are also suitable. Light sandy and heavy clayey soils are unsuitable. It is a hardy plant that thrives in cooler climates. In Indian plains, it is grown during rabi season as a cool season crop. It is a long day plant. Occurrence of strong winds and heavy rains during flowering is injurious. Oil yield is much reduced, if extreme heat coincide with crop maturity.

Propagation is through direct line sowing of seeds usually in September-October in plains and in March-April in temperate areas. If sowing is delayed, flowering occurs before proper attainment of vegetative growth and seed yield is considerably reduced. Seed rate is 5 kg/ha. Seeds remain viable for three years. Seed germination occurs within 7-9 days with more than 95% germination. First thinning is done one month after sowing when seedlings are about 15 cm tall to adjust spacing to 10 cm between plants to avoid over crowding. One month later, second thinning is done to adjust plant to plant spacing to about 23 cm. Irrigation and weeding are to be done on a need basis. FYM at 5–10 tonnes/ha is incorporated as basal dose. After second thinning and at blooming stage, application of 50 kg N/ha is beneficial.

Crop matures in five to six months. Quality of oil varies with maturity of crop and weather conditions prevailing at maturity period. A crop sown in mid October will be ready for harvest by end of April. For oil extraction from herb, harvesting is done when plant is of 105 days age, immediately after blooming period when seeds are immature.

For seed oil, harvesting should be done when seeds are fully mature but before shattering occurs. If harvested when seeds are fully developed but still green in colour, oil content is more (2.6-4.0%) and carvone content will be within pharmacoepial limits (43-63% w/w). At later stages, oil content drops to 1.7%. Harvesting should be done in morning hours when plants are damp with dew to minimize loss by shattering and volatilization. Seed yield is 5- 7 quintals per hectare.

Herbage yield is around 2.5 to 3 tonnes/ha which on distillation yields about 18-20 kg oil. It contains less than 40% usually 34-36% carvone. Dill seed oil with 20% or less of carvone content is preferred due to better flavour. Principal constituent of herb oil are phellandrene and limonene and that of seed oil is carvone. Herb oil from crop at advanced stage of growth contains more carvone and would assume odour of seed oil.

Indian dill herb yields 0.62% volatile oil on steam distillation. It usually contains 12-15% dihydrocarvone and 8.6% dillapiole. Indian dill seed oil content varies from 1.2- 3.5% and it contains 15.6–39.6% dillapiole. Oil has a high specific gravity due to presence of dillapiole. Dillapiole is a toxic substance, not found in European dill seed oil. Indian dill seed oil is not preferred in pharmacy due to high dillapiole content and low carvone content.

Pest and disease management

No much pest damage is reported in the crop. The fungus, *Erysiphe anethi* causes serious damage between seedling and flowering stage which can be controlled by copper fungicides. Root rot by *Fusarium* also causes considerable damage to crop.

Extraction and utilisation

Essential oil is extracted from aerial parts along with immature fruits and also from mature harvested fruits. Herbage and umbels should be dried only under shade. These oils differ in their composition, odour and flavour. Steam distillation takes 3-4 hours for oil extraction in fresh, immature herbage whereas 8-9 hours are required for mature and dried herbage.

Typical flavour of herb oil is due to its α-phellandrene content. The herb character predominates as long as oil contains less than 35% carvone. Oils of only 20% and less carvone content have the finest herb character. To retain herb character and also to prevent loss of more volatile terpenes, it should be distilled as fresh as possible or with minimum drying. Herb oil is preferred in food industry for flavouring and seasoning purposes. Oil of herb is colourless to brownish yellow. Oil content of fresh herbage varies from 0.60–2.84% at zero moisture and oil content of fresh leaves increases up to peak flowering and then declines.

Seed oil with its high carvone content resembles caraway seed oil. It is sweet, fruity, fresh and caraway like. It contains limonene 10%, terpinene 6%, dihydrocarvone 12%, carvone 34.5%, carveol 4%, dihydrocarveol 3.5% and isoeugenol 2.5%. Total ketone content expressed as carvone comes around 50 %. It is light yellow in colour, becoming dark in storage.

Oleoresins are produced primarily from dill seed, fortified with dill weed oil to more closely resemble the flavour character of whole herb. It is light amber to green in colour with 70 % volatile oil content (v/w).

Physico-chemical properties of oil

a. Dill seed oil

Appearance	Pale yellow to light brown coloured clear liquid
Specific gravity (25°C)	0.925 to 0.980
Optical rotation	40° to 58°
Refractive index at 20°C	1.486 to 1.495
Solubility 90% ethanol	0.5 v/v
Carbonyl content calculated as carvone	20 to 30%

b. Dill herb oil

Specific gravity (15°C)	0.878 to 0.908
Optical rotation	81° 10′ to 101° 4′
Refractive index at 20°C	1.4829 to 1.4855
Solubility in 80% ethanol	Soluble/soluble with cloudiness
Carvone content	25.6 to 42.8 %

Galangal

Scientific name: *Kaempferia galanga* L.; Syn: *K. sessilis* Koenig; *K. plantaginifolia* Salisb.
Family: Zingiberaceae
Vernacular names: *Sanskrit*: Karcurah; *Hindi*: Candramula; *Bengali*: Chandumula; *Malayalam*: Kacholam, Kachooram; *Tamil*: Kaccolam; *Kannada*: Kacora; *Telugu*: Candramula

Galangal is a rhizomatous perennial plant, rhizomes of which yield an essential oil. Oil is utilized in manufacture of perfumes and in curry flavouring. It is also employed in cosmetics, mouth washes, hair tonics and toiletries. Pungent, hot, sharp, bitter and aromatic rhizomes find an important place in indigenous medicine as stimulant, expectorant, diuretic and carminative. It promotes digestion and cures skin diseases, piles, phantom tumors, coughs, oedema, fever, epilepsy, spleenic disorders, wounds, asthma and rheumatism. Rhizomes are used for protecting clothes against insects and are eaten along with betel and arecanuts as a masticatory. Rhizomes and leaves are attached to necklaces and added to bath water for perfume.

Habitat and distribution

Plant is supposed to have originated in East Asia, most probably in Burma. It is widely distributed in Asia, Africa and Australia. It is grown in India, Burma, China, Nigeria, Mexico and other neighbouring countries. In India it is cultivated mainly in Kerala, Karnataka, Tamil Nadu and West Bengal.

The plant

Galangal is a pentaploid with x=11. Plant is a stemless perennial herb with tuberous, aromatic rootstock having fleshy cylindrical nonaromatic root fibres. Leaves are round to ovate, deltoid, acuminate, 10-12 ribbed, 6-12 cm long, 4.5-9.0 cm wide and horizontally spreading. Inflorescence is a scape with 6-12 flowers, figatious and fragrant, normally open successively and seldom set fruits.

Cultivation

K. galanga requires a warm humid climate. It thrives well up to an elevation of 1500 m. A well distributed annual rainfall of 1500-2500 mm during growing period and dry spells during land preparation and harvesting are ideal. Rich loamy soil with good drainage is suitable for cultivation of crop. Laterite soil with heavy application of organic matter is also suited. It is susceptible to waterlogging.

Plant is propagated by division of rhizomes. For planting mother rhizomes are better than finger rhizomes. Seed rhizomes are stored in cool dry place or in pits dug under shade. Smoking of rhizomes prior to planting is beneficial for better germination and establishment of sprouts. With receipt of pre-monsoon showers in May, land is ploughed and beds of 1-2 m width, 25 cm height and convenient length are taken and sprouted seeds are planted at 15-30 cm spacing. Seed rate is 500-750 kg/ha. Kacholam responds well to organic manuring. Application of 30 tonnes/ha of FYM or compost and mulching with leaves or straw at 15-20 tonnes/ha are recommended. Application of 50-75 kg each of N, P_2O_5 and K_2O in 2-3 splits is beneficial. Weeding is to be done 45 and 90 days after planting, followed by fertilizer application and earthing up. It is a shade loving plant. Growth is better in partial shade offering great potential for its cultivation as intercrop in coconut, arecanut, banana and other widely spaced perennial crops. Growth and rhizome yield (6.1 compared with 4.8 tonnes/ha) were higher when *K. galanga* was grown as intercrop in coconut compared with a sole crop. Essential oil and oleoresin contents were also higher in intercropped-rhizomes.

Crop is harvested 6-7 months after planting when leaves start drying up. Rhizomes are dug out, cleaned and washed to remove adhering soil particles.

Pest and disease management

Insect pests are not commonly reported in this crop. Leaf spot and rhizome rot diseases occur particularly during rainy months which could be controlled by drenching and spraying with 1% Bordeaux Mixture.

Extraction and utilisation

Clean rhizomes are sliced to circular pieces of uniform size and dried for 3-5 days. Sliced and dried rhizomes are marketed. Yield, on an average, is 5-8 tonnes/ha of fresh rhizomes which on drying yields 1.5-2 tonnes/ha of dry rhizomes. Driage varies from 23 to 28%. Sliced and dried rhizomes on steam distillation for 3-5 hours yield 2-3% of essential oil. Frothing is noticed during distillation due to presence of starch in rhizome.

Constituents of rhizome oil and root oil of *K. galanga* are similar in GLC studies. Ethyl cinnamate and ethyl p-methoxy cinnamate are major constituents. Volatile oil content of rhizome is higher than that of root.

Tuberous rhizome of kacholam contains an alkaloid, starch, gum, fatty matter with a fragrant liquid essential oil and a solid white crystalline substance and mineral matter. Rhizome possesses a camphoraceous odour with somewhat bitter aromatic taste resembling that of *Hedychium spicatum*. Essential oil has following properties.

Specific gravity at 30°C	0.8792-0.8914
Optical rotation at 30°C	-2° 36' to -4° 30'
Refractive index at 30°C	1.4173 to 1.4855
Acid value	0.5-1.3
Saponification value	99.5-109.0

Essential oil contains over 54 components of which major ones are ethyl-trans-p-methoxy cinnamate 16.5%, pentadecane 9%, 1, 8-cineole 5.7%, α-carene 3.3%, and borneol 2.7%. Terpenoid constituents amounted to 16.4%.

Greater Galangal

Scientific name: *Alpinia galanga* (Linn.) Willd.
Family: Zingiberaceae
Vernacular names: *Sanskrit*: Sugandhamula, Rasna; *Hindi*: Kulainjan; *Malayalam*: Aratta, Chittaratha; *Tamil*: Arattai; *Gujarathi*: Kolinjan; *Kannada*: Dumba-rasmi; *Marathi*: Kosht-Kulinjan; *Bengali*: Kulanjan; *Telugu*: Pedda-dumparashtram

Greater galangal, Java galangal or Siamese ginger is a perennial aromatic rhizomatous herb. This plant is cultivated for its rhizome in tropical areas of south and east India. Its essential oil is used in aromatherapy to relieve tension, improve mood or physical and emotional health and maintain health and beauty. It is especially useful for bronchial troubles and as a carminative. It is one of the ingredients of medicated "*Pan*" used for removing foul smell of mouth and getting relief in throat inflammation. In Ayurveda, "*Rasna-saptak-kwath*" and "*Rasna-adikamath*" are used as anti-inflammatory decoctions. In Unani, it is an ingredient of aphrodisiac preparations, "*Majun Mugawivi ma Mumsik*", "*Majun Samagh*", and antispasmodic nervine tonic "*Majun Chobchine*" and "*Lubab Motadil*". It is also used in "*Arq Pan*" as a cardiac stimulant and carminative.

Habitat and distribution

The Java galangal is native to Indonesia. It is distributed in India, China, Sri Lanka, Myanmar, Nepal and South East Asia. In India, it is very common in West Bengal, Bihar, Assam, Kerala, Karnataka and throughout the Western Ghats.

The plant

Greater Galangal is a perennial herb, about 2 m high with lower portion covered with smooth leaf sheaths. Leaves are broadly lanceolate, 30-60 cm long and 10-15 cm broad. Flowers are arranged in erect, terminal panicles, composed of numerous spreading dichotomous branches, each with two to six, pale greenish-white and faintly fragrant flowers. Fruits are 1.25 cm long, oblong, constricted in middle or even pear shaped, three sided and deep orange red in colour. Seeds are ash coloured, three angled, finely striated towards hilum. Both seeds and rhizomes have pungent aroma.

Alpinia Calcarata (Linn.) Willd is another species of genus with much medicinal importance. It is shorter in stature but stronger in aroma than *A. galanga*.

Cultivation

Siamese ginger comes up well in tropical climate with an annual rainfall ranging from 1500-3000 cm. It grows on a wide range of climates and soils. Well drained hilly areas and places of 1400 m height are good for its cultivation. Fertile red loams to forest soils are suitable. This is commercially propagated vegetatively by rhizomes. Field should be ploughed to a good tilth. All stones and pebbles should be removed. Organic manures at 10 tonnes/ha are applied during land preparation. Seedbeds of 1 m breadth, 2 m length and 15 cm height are prepared. Small pits are made in seedbeds and 5 cm long rhizomes are planted. Cover rhizome with FYM and mulch seed bed with leaves or straw. Optimum spacing is 30x20 cm under good fertility and 40x30 cm under poor fertility conditions. Fresh healthy disease free rhizome bits with at least one viable shoot bud is used for planting. Seed rate is 1000-1500 kg/ ha. It is irrigated immediately after planting. Carry out gap filling within one month, remove weeds two months after planting followed by top dressing, earthing up and mulching. Thereafter no weeding is required as crop smothers weeds. Apply fertilizers at 100:50:50 kg NPK/ha/year in 2-3 split doses. Application of biofertilizer *Azospirillum* at 10 kg/ha and cowpea green manuring *in situ* are beneficial for crop. This is also cultivated as an intercrop in coconut or rubber plantations. Though the crop can be harvested after 18 months, optimum stage of harvest for obtaining maximum rhizome and oil yield is 36-42 months after planting. Cut and remove shoot portion and carefully dig out rhizomes and roots. Harvesting is very arduous due to strong and extensive root ramification. Rhizomes are dug out after cutting top portions. Average yield is 10-15 tonnes of fresh rhizomes/ha and the driage is 25-30%.

Extraction and utilisation

Collected rhizomes are washed and cut into 5cm long pieces and dried in sun for 4 days before sale.

Rhizome contains tannins and flavonoids, some of which are identified as kaempferide, galangin and alpinin. Seeds contain 1′-acetoxychavicol acetate and 1′-acetoxy eugenol acetate, caryophyllenols I and II, n-pentadecane, 7-heptadecane and fatty acid methyl esters. Rhizomes yield essential oil containing methyl cinnamate, cineole and α-pinene and sesquiterpenoids. Fresh rhizome yielded 18 monoterpenoids of which α-pinene, β-pinene and limonene as major compounds and 17 oxygen containing monoterpenoids with cineole, terpinen-4-o1, and α-terpineol as minor compounds. Fresh rhizomes on steam distillation for 3-5 hours give 0.22% essential oil. Oil recovery on dry weight basis is 0.93%. Root is also a significant contributor of essential oil.

Rhizomes are bitter, acrid, thermogenic, aromatic, nervine tonic, stimulant, revulsive, carminative, stomachic, disinfectant, aphrodisiac, expectorant, broncho-dilator, antifungal, febrifuge, antiinflammatory and tonic. Rhizome is central nervous system active, diuretic, hypothermic. Seed is antiulcerative. Rhizome spray in ether, over a space showed high knock down values against houseflies. Alcohol (50%) extract of rhizome is anti-amphetaminic. Unani physicians consider it good against impotence.

Hops

Scientific name: *Humulus lupulus* L.

Family: Cannabinaceae

Vernacular names: *English*: Hop; *French*: Houblon; *German*: Hopfen; Hungarian: Komlo; *Italian*: Lopolo; *Persian*: Hymel.

Hops is the dried conical catkins (strobile) of female plants of *H. lupulus*. Cones bear glandular trichomes at base of each scale, surrounded by a pale yellow granular powder, called lupulin. Lupulin grains at maturity are opaque resembling "flower of sulphur" and can easily be shaken from flowers. Lupulin is an important raw material for brewing industry throughout world. Lupulin contains resins and essential oil, imparting to beer, ale, stout, porter and other malt beverages their characteristic aroma and bitter flavour, considered desirable by consumers. Simple decoction and extract of hops are used as a tonic. Hop concrete and hop absolute are used for flavouring. Hop absolute introduces a most interesting naturalness to apple flavours and pine apple. In perfumes, it offers a bitter green note which is interesting in pine, hyacinth, citrus and in aldehydic perfumes.

Habitat and distribution

A Peruvian mission introduced hops to Lahaul and Spiti valley and Keylong areas in 1862. Although its cultivation subsequently spread over a sizable area eventually cultivation was given up. Systematic hops cultivation was started in India by British government in Kashmir during reign of Maharaja Pratap Singh, some time during later part of 19th century. A plantation of about 40 ha was established in Doaba, about 41 km from Srinagar with a production capacity of about 22 tonnes of hops annually. However, yielding to stiff competition from hops imported from U.S.A and Europe, Kashmir hopyard was closed down during second decade of 20th century. Although hops was also tried in Lal bagh Gardens at Bangalore in 1954, commercial cultivation was not ventured because of unfavorable weather conditions.

Today, hops is cultivated on commercial scale in Argentina, Australia, Austria, Belgium, Canada, Czekoslovakia, England, France, Germany, Hungary, Japan, Korea, New Zealand, Poland, Romania, South Africa, Spain, U.S.S.R, Yogoslavia and U.S.A.

The plant

Family Cannabinaceae in which hops is included consists of only two genera -*Humulus* and *Cannabis*. *Humulus* is represented by two species; hops (*H. lupulus* L) and the Japanese hops (*H. japonicus* Sieb). All cultivated varieties belong to *H. lupulus*.

Hops are perennial in nature, climbing in habit and grows to a height of 8-12 m along poles or any other standard. Root is stout and perennial. Stem that arises from it every year is of a twining nature, reaching a great length, flexible and very tough angled and prickly, with a tenacious fibre. Hops is dioecious. Male flowers are much branched cymes in panicles which grow from axils of leaves of lateral branches. Individual flower is 0.6 cm in diameter and 0.2 cm in depth. Female inflorescence is formed during May–July. It is borne on axils of leaves of lateral branches with round terminal buds. It is green in colour and carries alternate pairs of small bracts. Flower is simply a cup shaped perianth enclosing an ovary, provided with two comparatively long papillated stigmas. It is borne on thin short pedicel arising from a shallow concave depression at end of protuberance from axis. Whole inflorescence is a brush-like structure called burr. Ovary is nearly 1 mm long and stigmas are almost 3 mm long.

After pollination, stigma quickly dies and axis elongates. Bracts and bractioles quickly enlarge and inflorescence assumes a cone-like appearance. Mature cones retain structure of burr. Golden yellow resin glands containing valuable brewing material "lupulin" are clustered on outer surface of lower part of bractioles and over entire surface of perianth. Lupulin gland is developed from epidermal cells which grow out from surface of bractioles and divide to form a multicellular, single layer cells in the form of a cup. Cups develop by the time burr is fully developed and secretion of resin commences before cone starts to grow out.

Cultivation

Rich alluvial or deep fine sandy loam and medium loam soils are the most suitable for hop cultivation. Heavy, wet and strongly alkaline or saline soils should be avoided. The pH of soil may vary from 6.0 to 8.0. Hops are propagated by seedlings and from root cuttings; but seed propagated plants take longer time to come to maturity than ones raised from cuttings. Commercially hops are propagated through cuttings as vegetatively propagated crop alone assures uniform maturity and consistent yield. Female plants alone are propagated as only pistillate flowers bear lupulin glands. Further, these glands develop without fertilization. Land for hop is well prepared by repeated ploughing. Cuttings of 15 to 20 cm are collected from young disease free plants. Two selected cuttings are planted vertically in each hole at a depth of 20-25 cm at a spacing of 1.5x1.5 m. Before planting, cuttings should be dipped in 0.3% cuman L or 0.3% benlate for 5 to 10 min. About 3000 plants can be accommodated in a hectare of land.

Being weak stemmed, plants are provided with mechanical support. Staking with poles, trailing on trellice made with coir rope or iron wire are a few of the methods followed by hops cultivators. Recently high wire trellice composed of 5-6 m high angle iron poles with iron wire stretched between them is widely used. About 4 to 5 hoeings are needed during the year to assure weed control and soil aeration. For chemical control of broad leaved weeds, 2,4-D @ 2.5-3 kg/ha and for grasses gramaxone at 2.5 l/ha in 500 l of water are used. After each hoeing, earthing up to 60 cm height is essential. Application of fertilizer nitrogen, phosphate and potash is essential to assure good yields of hops. On an average, 150 kg N, 150 kg P_2O_5 and 100 kg K_2O per ha are applied in five equal splits in a year. Zinc and copper stimulate growth and increase cone yield significantly.

After ripening of bine, it dies back to root stock and at this stage, bines are scissor pruned to remove dead and diseased plant parts. Pruning is aimed to limit number of shoots to be formed in next season and to suppress number of unproductive shoots. Pruning is done thrice a year in April, May and June and operation is followed by earthing up.

Picking of hops is done at full maturity when flowers are yellowish green in colour, brittle in texture and possess an agreeable aroma. Generally in Kashmir, flowers mature from last week of August and continue up to end of September. Picking should be done within 10 days of ripening. If allowed to overripe, cones are more fragile and also tend to lose some of their lupulin, which may be shaken out by wind. In Kashmir, picking is done in September-October by manual labour. Time is ideal in Kashmir, since there is no danger of damage to cones by rains. Harvesting should be done carefully to exclude leaves and stem portions as latter contains several constituents undesirable to brewing industry.

Extraction and utilisation

Freshly picked hops contain 65-80% moisture which must be reduced to about 12% to prevent quality deterioration during storage and transportation. This is done by careful slow drying. Dried hops may be heaped for 10-14 days before they are baled and shipped to brewing houses.

Alternatively, essential oil of hops can be obtained by steam distillation of freshly dried cones. Old hops results in lesser amount of essential oil. Great care should be bestowed during distillation to control distillation rate as rate of distillation influences oil quality; very high or low rates of distillation results in poor quality of oil. Yield of oil varies from 0.2 to 0.8% depending upon age of dried hops. Average herb yield in first year of plantation is 700 kg dried hops per hectare which increases to around 1400 kg/ha in third year. Thereafter it remains steady up to 15 years.

Hop concrete is produced by solvent extraction with either gasoline or benzene. The concrete is in turn extracted with alcohol and solvent removed under reduced pressure to yield hop absolute.

Physico-chemical properties of oil

Specific gravity (15°C)	0.855-0.899
Optical rotation	-1° to 2°27′
Refractive index (20°C)	1.4850 to 1.4968
Acid no.	0.5 to 10.00
Ester no.	13 to 40
Ester no. after acetylation	18-46
Solubility	Very difficultly soluble in alcohol.

Oil contains following chemical constituents; myrcene : 30-50%, humulene: 15-25%, esters of myrecenol : 20-40% and traces of dipentene, linalool, farnesene, β-caryophyllene and free and esterified acids such as formic, acetic, buyric, caprylic, pelargonic and capric acids, caryophyllene and humulene.

Jasmine

Scientific name: *Jasminum grandiflorum* L.; Syn: *J. officinale* (Linn.)
Family: Oleaceae
Vernacular names: *Sanskrit*: Jati; *Hindi*: Jati; *Bengali*: Jati; *Malayalam*: Pichakam; *Tamil*: Malligae; *Kannada*: Gundumalligae; *Telugu*: Malati; *Gujarathi*: Ghambeli; *Punjabi*: Jati

Jasmines are a group of shrubs commercially grown for their fragrant flowers and essential oil production. Bulk of flowers is used as such in garlands and decorative bunches for religious offerings and a small quantity for production of oils and attars. Jasmine concrete and absolute are used in high grade perfumes, ranking next to rose in order of importance. Jasmine oil blends with every floral scent and extensively used as an important perfumery item throughout world. Almost all high quality perfumes contain at least a small amount of jasmine oil. Absolute, though expensive, also blends with any floral scent imparting smoothness and elegance to perfume composition.

Habitat and distribution

The term Jasmine is probably derived from Persian word *Yasmin* meaning *fragrance* which is adopted in Arabic as *Yasym* given to jasmine flowers. Of many species of jasmine, commercially cultivated species, namely *J. grandiflorum* is a native of Kashmir, Afghanistan and Iran, *J. sambac* a native of South India and *J. auriculatum* is a native of South and Central India. A study of ancient Tamil literature of "Sangam" period (500 BC to 200 AD) revealed that all these three species are mentioned in those ancient works and therefore South India, would have been an important centre of origin of *Jasminum* species. Jasmines are widely found in warm parts of Europe, Asia, Africa and Pacific regions but almost absent in America. Annual production of jasmine concrete is more than 15 tonnes, the largest producer being Egypt followed by Morocco, India, Italy, France and China. Jasmine oil has great export potential. Oil is mainly exported to Europe, Srilanka, Singapore, Malasia and some Arabian conutries.

The plant

Jasmines are climbing, trailing and erect shrubs. There are both evergreen and deciduous species. Leaves are opposite or alternate, simple, trifoliate or pinnate, leaflets are entire. Flowers are white, yellow or rarely reddish, sometimes solitary, more often in cymose clusters of three to many, usually fragrant. Corolla is tubular with four to nine lobes. There are two stamens; ovary is two loculed with 1-4 erect ovules. Fruit is a berry and black in colour.

Though more than 2000 species are known, about 4 species, viz. *J. grandiflorum, J. auriculatam, J. sambac* and *J.odorotissimum* bear fragrant flowers, of which three species, viz. *J. grandiflorum* L., *J. sambac* Ait. and *J. auriculatum* Vahl. are commercially cultivated and only *J. grandiflorum* is grown for use in perfume. About 22 varieties are under cultivation in different parts of India.

Jasminum grandiflorum is a twining or nearly erect growing shrub; branches ribbed, drooping, annular; leaves opposite, imparipinnately compound, rachis flattened or winged; leaflets 5-7, elliptic, round or oval; flowers borne on axillary or terminal cymes longer than leaves, white, often tinged purple on outside, fragrant; bracts ovate to spathulate, oblong, foliaceous; calyx glabrous, 5-lobbed, subulate; corolla 5-lobbed, star shaped, elliptic or obovate; corolla tube encloses 2 stamens borne on short slender filaments; ovary bicarpellary; fruit berry. The common varieties of this species are Co-1 Pitchi, Co-2 Pitchi, Thimmapuram and Lucknow.

Jasminum sambac is a dwarf spreading bushy shrub, 0.5-1.0 m high with attractive glabrous leaves producing attractive, white, sweet scented flowers in great profusion in hot season. It is the most ideal species for cultivation in Kerala. Gundumalli,Motia,Virupakshi, Sujimalli, Madanabanam and Ramabanam are a few important varieties of *J. sambac.*

Jasminum auriculatum is a twining scandent shrub growing to 5-7 m with small opposite leaves; flowers white and sweet smelling with calyx fine notched having round firm glandular process on outside, light weight; around 26,000 flowers weighing a kilogram. Co-1 Mulla, Co-2 Mulla, Long Point, Long Round, Short Point and Short Round are common varieties of *J. auriculatum.*

Cultivation

Jasmines are sun loving plants and prefer warm humid climate for successful growth. They perform well at elevations ranging from 600 to 1200 m. Areas having a warm summer and mild winter with sun almost throughout year are considered the best. The best time for planting is July and August and from end of January to February in north India and any time between July to December in south India. Jasmines can be grown on a wide range of soils, but well drained rich sandy loam to clay loam soils with a pH of 6.5-7.5 are the best suited. Soil should be thoroughly prepared and kept free from weeds.

Layers and cuttings can be used for propagation. Improved varieties of *J. grandiflorum* ('Jaji mallige' and 'Ajjige') and *J. auriculatum* ('Vasantha mallige'

'Parimullai' and 'CO -1 mullai') are used for commercial cultivation. Cuttings for planting should be 20-25 cm long with 3-4 eyes and are dipped in seradix-B (or a solution of 4000 ppm of IBA) and planted under intermittent mist for rooting during January to March. Almost 90% rooting is achieved and cuttings will be ready for transplanting in 4-5 months. After plouhing land well, pits of about 40 x 40 x 40 cm size are taken and filled with top soil, cowdung and compost. Plants are spaced at 1 m in rows, 1.5 m apart. Irrigation is given if soil moisture are inadequate.

Jasmine plant needs 15-30 kg FYM, 60-120 g N, 120-240 g P_2O_5 and 120-240 g K_2O/plant/year, given in 3-4 split doses. Pruning is needed to get high flower yield and to keep bushes to manageable size. Plants pruned between December and June produce maximum number of branches and the highest yield of flowers. Shoots are cut at 30 cm height. Diseased leaves and dry shoots are also removed. After pruning, soil around plant should be stirred upto a depth of 15 cm and repeated every 2-3 months. Field should be kept clean by removing weeds as and when necessary.

Plant flowers from second year of planting. Flowering period ranges between April and May and from August to November. In Egypt, plant flowers almost 10 months in a year. Harvesting is done during early morning because flowers contain maximum perfume at this time. Flowers gathered at noon and in afternoon yield lesser flower oil than those collected very early in the day. Warm weather and ample sunshine yield a crop of heavily scented flowers than in cool or rainy weather. Annual yields of flowers are reported to be 750-1000 kg/ha in India, 2000-4000 kg in France, 4500-5500 kg in Sicily and Italy and upto 6000 kg/ha in Morocco and Egypt. A jasmine plantation gives economic yield for 10-15 years after which crop is removed and crop rotation followed for some years before establishing a new jasmine plantation.

Pest and disease management

Pests

Bud worm: Small green caterpillars of insect feeds on flower petals and bore into buds resulting in fading of colour destruction of buds due to excreta.

Blossom midge: Mosquito like adult midges lay eggs at base of flower buds. Maggots enter buds and feed from within.Buds get discoloured and dry off.

Flower thrips: Adults and nymphs of thrips feed on sap of buds and flowers. Affected flowers turn violet and dry off.

Gallery worm: Caterpillars of insect fold together leaves, buds and growing tips of branches and feed from within. If damage are severe, plants get stunted and die off.

Leaf thrips: Leaf thrips are seen in large numbers on lower surface of leaves and feed on leaf sap. Due to infestation, leaves become leathery and yellowish and drop off.

White flies: Adults and nymphs of white flies suck sap from lower surface of leaves and as a result, leaves become yellowish and pale.

Floral mites: Small yellowish mites infest floral buds. Infested buds do not open and get crinkled. Size of the buds will be reduced due to infestation.

Foliar mites: Mites infest in colonies on lower surface of leaves. Infested leaves become yellow and drop off.

Lace wing bugs: Adults and nymphs of these bugs feed on leaf sap from lower surface. Yellowish patches develop on leaf surface and later they dry and drop.

Grass hoppers: Grass hoppers feed on leaf lamina but generally not a serious problem.

Stem borer: Larva bore into stem and affected stem dries off.

Control of insect pests: Against caterpillars and leaf feeding insects, quinalphos EC at 2 ml/l, carbaryl WP at 4 g/l or hostathion EC at 2 ml/l shall be sprayed on need basis. Sap sucking insects can be controlled by dimethoate EC at 2 ml/l, phosphamidon EC at 0.5 ml/l, imidachloprid EC at 0.5 ml/l or acephate SP at 1.5 ml/l. Application of dicofol 2 ml/l, wettable sulphur 3 g/l or ethion 1 ml/l are effective against mites attacking jasmine.

Diseases

Leaf spot: Leaf spots appear on leaf surface and in severe cases, defoliation occurs. In initial stages of infection, affected leaves can be collected and destroyed. In case of severity, mancozeb 2-4 g/l or carbendazim 1 g/l or benomyl 2 g/l or captafol 3 g/l are effective.

Die back and rust: Symptoms of damage are drying of leaves, breaking of skin of branches and drying of entire branches. Removal and destruction of dried branches and spraying of fungicides such as mancozeb 2-4 g/l or carbendazim 1 g/l or benomyl 2 g/l or captafol 3 g/l are remedial measures.

Sootty mould: Black mould appears on leaf surfaces and reduces photosynthesis thereby affecting growth of plants. Spraying of diluted starch solution and 1% BM are effective against the problem.

Bacterial wilt: Plants get wilted in a week and die off by bacterial infection. Removal and destruction of affected plants and application of copper hydrate at 1 g/l or oxytetracyclin compounds (500 ppm) soil drenching are effective against disease.

Extraction and utilisation

Essential oil in flowers is extracted through enfleurage which is widely used for production of jasmine attars in India. In this method, seeds of sesame or til (*Sesamum indicum* L.) are first soaked in water to remove their covering and then dried in sun. Fresh jasmine flowers and dehusked sesame seeds are spread in thin layers, one above other, for 10-12 hours daily. Exhausted blossom is replaced by fresh flowers and this process is repeated for 5-7 days till all dehusked seeds are saturated with perfume. One kg seed can extract perfume from 3 kg flowers. Perfumed seeds are distilled and vapours of jasmine are absorbed into sandal wood oil for production of attars. Solvent extraction, with petroleum ether or hexane, recovers practically all odorous constituents. Solvent is recovered by vacuum distillation and residue constitutes concrete, purified by extraction with 95% alcohol, whereby jasmine absolute is obtained. Usual yields are 0.30-0.35% concrete and 45-55% absolute. Annual yield of concrete is 4-5 kg/ha. Jasmine oil is also separated from jasmine concrete by liquid carbon dioxide extraction method.

Jasmine concrete is a yellowish brown waxy mass with characteristic odour of jasmine flowers. It has melting point 50-51°C, congealing point 54-55°C, acid value 0.23-0.27 and it is partly soluble in 95% alcohol. Approximate composition of jasmine flower oil obtained by enfleurage is benzyl acetate 65.0%, δ-linalool 15.5%, linalyl acetate 7.5%, benzyl alcohol 6.0%, jasmone 3.0%, indole 2.5% and methyl anthranilate 0.5%. Benzyl benzoate, geraniol, nerol, terpineol, farnesol, nerolidol and *p*-cresol are also present in traces. Jasmine absolute is a viscous clear yellowish brown liquid possessing a delicate odour of fresh jasmine flowers. Absolute contains many of the above compounds.

Juniper

Scientific name : *Juniperus communis (Linn.)*; Syn: *J. sibirica, J. nana*
Family : Cupressaceae
Vernacular names: *Sanskrit*: Vapusha; *Hindi*: Aaraar; *Bengali*: Havusha; *Punjabi*: Petthri; *Arabic*: Abhal

The name *Juniperus communis* originated from latin words *Juniperus* meaning "juniper" and *communis* meaning, "common". It is widely known as common juniper. Other common names include Dwarf Juniper, Mountain Common Juniper, Old Field Common Juniper and Prostrate Juniper.

Junipers are a group of high altitude plants valued for their volatile oils. The most important of junipers is *J. communis*, which is renowned for its seeds. Seeds have high medicinal value and are also aromatic. Oil obtained from non-fermented berries of juniper is bitter with a characteristic odour. Oil is very expensive due to difficulties and rarity of collection of ripened fruits. It is used largely in compounded gin flavours, liquors and candies, ice creams, baked foods and chewing gums. It is used in fine fragrances for its unique unconventional fantasy notes.

Oil of Juniper is given as a diuretic, stomachic and carminative in indigestion, flatulence and diseases of kidney and bladder. Oil is also a local stimulant. They are used in disorders of genital tract, gonorrhea, gleet, and leucorrhea and in cutaneous diseases. Oil has aromatherapeutic property of local stimulant and in rheumatic pain of joint or muscles. Bulk of the commercial oil is obtained as a byproduct during distillation of alcoholic beverages. Oil, however is inferior to that obtained by direct distillation of fruits as it is partly deprived of natural oxygenated odoriferous compounds.

Juniper is an important spice in many European cuisines, especially in Alpine regions where juniper grows abundantly. It is the only example of a spice in botanic group of *coniferae* and also one of a few examples of spices from cold climatic regions, though the best quality stems are from southern European countries. Although juniper berries are harmless for healthy people, their massive use is discouraged for people with kidney weakness and pregnant women.

Habitat and distribution

Juniper is adapted for medium and high elevations in India. It prefers sunny open situations for a better growth. It is perhaps the most widely distributed tree in world. Plant grows well in sandy and loamy, moderately moist soils. It is found in typically dry, rocky, wooded hillsides or exposed slopes and in a variety of soil types including acidic and calcareous sands, loams, or marls, but grows well even in rather dry, rocky and gravelly ground. Juniper is widely distributed throughout the northern region of Asia and North America. Juniper tree is found in western Himalayas, Kumaon and Kurrna valleys at an altitude of 3500-4000 m. Generally killed or seriously damaged by fire; it is probable that seeds protected by overlying layers of soil survive at least some fires. Common juniper produces abundance of long viable seeds and after low-intensity fires, a few seeds germinate. A relatively long germination period and relatively poor germination rates contribute to slow post fire reestablishment on many sites.

The plant

This coniferous shrub or columnar tree, grows as a low, mat-forming shrub 0.75 -1.75 m tall and 2-4 m across. Leaves are in whorls of three, spreading, 1-1.25 cm long pungent, whitish above, convex or obtusely keeled beneath with a more or less prominent cushion on branchlets, persistent 3-4 years. Younger leaves tend to be more needlelike whereas mature leaves are scale like. Twigs yellowish or green when young, turn brown and harden with age. Bark thin, shreddy or scaly, often exfoliating into thin strips. Male and female flowers are borne on separate trees. Flowers axillary, supported by small imbricating bracts, male catkins ovoid, yellow antheroferous scales broad-ovate, acuminate and female flowers are small, resembling leaf buds. Berries blue black, 1.5-2 cm in diameter, tips of carpellary scales visible at apex, pulp sweet, resinous, seed generally 2-8. Fruits are berrylike; red, ripening to a glaucous blue-black colour. Fruits are ovoid to ellipsoid and contain one to three seeds. Fruit generally remains on plant for at least 2 years, with dispersal in August of second season.

Cultivation

Juniper can be propagated either by seeds, cuttings, layers or grafting. Seeds of most juniper require a specific period of rest after ripening. Generally, germination rate of seeds that are not fully-ripened is only around 1%. High temperatures, alternating temperatures, freezing and thawing, removal of seed coat, or application of hydrogen peroxide, dilute acids, carbon dioxide, or light had little influence on germination of juniper seeds. Juniper seeds have a semi permeable and thick seed coat with a dormant embryo. Seeds of common juniper require a long germination period. Seed dispersal is by gravity, water, birds, or mammals. Digestive processes apparently do not harm most juniper seeds and may actually enhance germination. Birds are the most important dispersal agents. Seeds are viable for several years if stored in cool, dry place. When sown, they often take a year to germinate, though sometimes they may be vegetative in a few years. Flowering takes place in April-May and fruits ripen in August-September of second year. Since juniper berries take two or three years to ripen, blue and green berries occur on same plant.

Extraction and utilisation

All parts of the plant contain volatile oil. Commercial oil of Juniper is obtained cheaply from ripe fruits. Steam distillation of ripe fruits yield 0.8-1.6% of essential oil. Oil from green unripe fruit is inferior; in over ripe fruits, oil changes into a resin. Needles and terminal twigs yield 0.15-0.18% of bright yellow oil. In India, leaf oil is acceptable to industry.

Physico-chemical properties of oil

Appearance : Colourless to pale yellow bright mobile liquid

Odour : Oil has fresh terpenic characteristic odour followed by dry spicy herbal woody base notes.

Specific gravity (27°C)	0.9180
Refractive index (27°C)	1.4820
Optical rotation	20° 8'
Acid value	4.7
Ester value.	20.5

Steam distilled juniper berry oil contains 24 components. Oil contains high percentage of terpene hydrocarbons (86.7%), non-oxygenated monoterpenes, α-pinene and myrcene being the major components (36.9 and 30.9 respectively) followed by limonene.

Needle oil contains 41 components accounting 94% of total oil. Major constituent is sabinene (27.5%) followed by elemol (15.8%), limonene (13.9%), bornyl acetate (7.2%), myrcene (5.3%), α-pinene (4.0%) and terpinen–4–ol (2.5%).

Kewda

Scientific name: *Pandanus fascicularis* Lam.; Syn: *P. odoratissimus* Linn.
Family : Pandanaceae
Vernacular names: *Sanskrit*: Ketaki, Soochikapushpa, Legupushpa; *Hindi*: Kedgi, Kevda; *Malayalam*: Kaitha, Kaida, Thala; *English*: Fragrant screwed pine, Umbrella tree, Screw pine, Screw tree; *Kannada*: Kedige; *Tamil*: Thazhai, Talai; *Marathi*: Kewda; *Telungu*: Mughlepuve, gajangi, ketki; *Gujarati*: Kewoda

Kewda or Screw pine is a shrub whose fragrant male spadices yield an oil called kewda oil of commerce. Fragrant smell is emitted by stamen and tender white spathes covering them. Flowers are processed to obtain essential oil (rooh), attars and kewda water. Oil contains methyl ether of phenyl alcohol (66.68%) which gives characteristic aroma to spadices. The 'kewda attar' is among ancient perfumery articles of India. It blends well with almost all types of perfumes and is used for scenting oils, tobacco, cloths, lotions, soaps and agarbatti. Kewda water, 'rooh' and 'attar' are used for flavouring various foods, sweets, syrups and soft drinks. They are quite popular for using in bath waters due to their oriental fragrance. Main use of kewda water is in pan masala and chewing tobacco industry. These two products consume nearly 80% of all attar produced. In recent years there is demand for this traditional Indian perfume abroad and consequently export is increasing. Juice of inflorescence is useful in rheumatic arthritis in animals. In addition, plant is considered a good soil binder.

Habitat and distribution

Kewda needs tropical climate with good rainfall for favourable growth. plant cannot withstand frost. It is found in tropical region extending from western Africa to Eastern Polynesia. Plants grow generally on banks of rivers, canals, fields and ponds. In wild state kewda grows along coasts producing a thick belt of unpenetrable vegetation above the high water mark. Though it grows near water source, it is hardy and drought resistant. In India, luxuriant growth of plant is seen along coastal belt of Ganjan district of Orissa and to some extent in neighbouring Srikakulam district of Andhra Pradesh. It is also found in Tamil Nadu, Kerala, Gujarat and some parts of U.P. In Ganjan district, kewda population is marked in about 5000 hectares in both forest and cultivated land. About 90% of country's Kewda oil is produced in Ganjan district. An estimated 40 million kewda flowers valued at Rs.200 million are processed annually in this region.

The plant

Kewda is a densely branched shrub or small tree with aerial stilt roots and stem up to 3-6 m height. Leaves are closely spirally arranged, caudate, acuminate, glossy green, 0.9–1.5 meters long, margins with forward pointed spines along edges and in midribs. Plants are dioecious. Individual trees are unable to fertilize themselves but apomixes in few females is suspected. They are unisexual, bearing male and female flowers in separate trees. Female flowers do not have any characteristic smell but bear aggregate fruits comprising of numerous ovaries that develop into drupes. Spadix of female flower is solitary, 5 cm in diameter. Spadix of male flowers are 15-40 cm long with numerous subsessile cylindrical spikes. Male inflorescences are valued for fragrant smell emitted by stamen and tender white spathes covering them and valuable attar obtained from them. Plant takes five years to flower and continue bearing for 40-50 years thereafter. Flowering will be maximum between 15 and 25 years. Flowering of kewda takes place throughout the year but 70% flowering occurs in rainy season (June to September).

Cultivation

It is a hardy plant and in nature, grows in waste lands, marshy lands and on bunds. However, for commercial cultivation, it prefers fertile and well drained soil.

Kewda is propagated through 20-30 cm long cuttings made from 8-10 cm thick stems. Aerial roots are also used in propagation. While making cuttings, non- flowering shoots and old stumps are preferred. They are planted in specially prepared sand beds under shade for sprouting. Beds are regularly watered and in about 45-50 days when they start sprouting, they are ready for transplanting into main field.

Before planting, main field is thoroughly prepared by 2-3 ploughings and harrowings. Pits of 60 cm^3 are dug at a spacing of 1x2 m using paired row method of planting in such a way that a pair of row spaced at 1x1m is created at every 2 meters of spacing. They are filled with mixture of soil and manure at least 2-3 weeks earlier to planting so that it settles down. Depending upon rainfall, planting is done from June-August.

A five year old plant attains an average height of 170cm with a stem circumference of 50 cm and with 20-25 aerial roots and on an average bears four flowers. At 20-25 years age, about 24 flowers are produced per plant. Flowers are plucked by hand or by pulling with a hooked stick. Collection by cutting down flowering stalk is not recommended as it may damage floral buds.

Extraction and utilisation

Flowers are harvested in early morning and immediately transported to distillation units preferably before 9.00 am for distillation of oil. Longer storage results in deterioration of product quality. Kewda flowers are processed by traditional water distillation. Flowers are distilled for two types of products. Essential oil from flowers obtained by water distillation is known as 'Rooh' of kewda or absolute. Other product being 'attar' obtained by absorbing hot vapours from kewda flowers into sandalwood oil. However, due to shortage in availability of sandalwood oil and high price, cheaper attars are produced by absorbing vapours in purified paraffin oil or dioctyl phthalate base.

Essential oil or 'Rooh' is produced during rainy season (peak season). When ambient temperature is low, a better recovery of oil is obtained. During hot days or when distillation is started late in morning, recoveries are unsatisfactory and uneconomical. During these days, flowers are processed for attar making only.

Physico-chemical properties of oil

Appearance	Thin low viscous clear liquid
Color	Colorless to pale yellow
Odor	Strong and characteristic
Relative density at 27°C	0.97
Refractive index at 27°C	1.493-1.495
Optical rotation	2.3° to 4.4°
Hydroxyl value	1.5-2.5
Methoxyl value	15.2-18.3
Solubility in aqueous alcohol	Completely soluble in two columns of 70% alcohol.

The chemical constituents of oil are β-phenyl ethyl methyl ether (65-80%), terpenen-4-ol (10-20%), ρ-cymene (0.3-3.3%), α-pinene (0.5-0.6%), β-pinene (0.05-1.8%), γ-terpinene (0.05-2.5%), β-phenethyl ethyl alcohol (0.1-1.0%), α-terpeneol (1.1-3.2%). Other minor components detected are limonene, piperitone, linalool and carvenone.

Lippia

Scientific name: *Lippia alba* Mill NE Br.
Family: Verbenaceae
Vernacular names: *Hindi*: Kavach

Lippia is a herbaceous plant of South American origin. Plant is aromatic and its leaves contain most of its essential oils. Plant existed in different chemotypes, one having citral as main constituent, others with linalool, camphor and 1,8 cineole as major constituents in their essential oils. Essential oil of Lippia is used in traditional system of medicine and also for control of pest infestation of stored food grains. Oil of *L. alba* has analgesic, anti-inflammatory and fungi-toxic properties.

Habitat and distribution

Lippia consists of about 200 species of herbs, shrubs and small trees, widely distributed in topical to semi- temperate areas of America, Africa and Asia. In India, plants of Lippia species are mostly found growing as weed in wet conditions along muddy river banks in Assam, Bihar, Orissa, Madhya Pradesh and Tamil Nadu.

The plant

It is a spreading shrub growing up to a metre in height, much branched and branches straggling upto 2 m. Stem is hard and brownish green. Leaf is 3-12 cm in length, 2-6 cm in width, ovate–oblong or elliptic oblong, apex acute, margin finely crenate serrate, base truncate to cuneate and softly strigose. Flowers are sessile arranged in axillary dense capitate spikes, corolla light purple with yellow throat. Fruits are globose, splitting into two one-sided pyrenes.

A new distinct variety, christened as Kavach of *L.alba* was developed through selection from an accession LAC-2. This reproductively unfit and vegetatively propagated variety yielded an essential oil rich in linalool (65%), one of the most useful ingredients of perfumes and precursor of vitamin E. Whole plant herb of Kavach contained 0.2% oil content, while a higher concentration of 0.8% was in leaves and flowers. The herb and oil yield over single harvest were recorded as 18.5 tonnes/ha and 37 kg/ha, respectively.

Cultivation

It is propagated through vegetative cuttings. Studies on cutting propagation of *L. alba* with different types of semi-hardwood cuttings (apical and middle with 0, 2 and 4 leaves), different hardwood cutting lengths (5, 10, 15 and 20 cm) and different substrates (carbonized rice hulls, vermiculite, soil and Plantavax) showed that all types of cuttings showed high rates of rooting. Middle cuttings with 4 leaves showed the highest root development, while leafless cuttings showed the lowest. Cuttings with 2 leaves also showed good development, easy manipulation and high yield. Substrates did not affect rooting percentage, but the largest root mass was obtained in carbonized rice hulls. Increase in hardwood cutting length caused a linear growth in all variables. It was established that *L. alba* can be propagated with semi-hardwood cuttings with one pair of leaves or with hardwood cuttings with 20 cm length, in porous substrate and without intermittent mist.

In another study, effects of cutting position (node or internode), cutting length (<25 cm and >25 cm), and cutting diameter (<5.0 mm and >5.0 mm) on rooting of *L. alba* cuttings were examined at Paulista Statal University, Brazil. Cutting position and length did not affect rooting or leaf volume. However, cutting diameter did significantly affect rooting and leaf production, thicker cuttings giving better results.

Rooted single plants of *L. alba* var. kavach are transplanted in field. Plants are irrigated to maintain soil moisture. Application of organic manures and inorganic fertilizers stimulates growth and yield. Variety Kavach can be harvested three times a year and yields more than 100 kg essential oil per hectare and thus could serve as a rich source of linalool and appears especially suitable for cultivation on slopes, river banks, ridges and ravines to check soil erosion.

Extraction and utilisation

Fresh herb is hydrodistilled for 3-4 hours to obtain essential oil. Alternatively, it is extracted with acetone which is then evaporated in vacuum in a rotary evaporator.

Major constituents of oil are geraniol (15.57%), an unreasolvable mixture of myrthenol and myrthenal (9.89%), nerol (9.44%), geraniol (7.36%), 2,6-octadien 1-01-3,7–dimethyl acetate (6.87%), 1–octen-3-d (4.60%), 6-methyl–5-hepten-2-one (4.60%), citronellol (2.63%), linalool (2.20%), 3-pinen–2-d (2.19%), b-myrcene (1.49%), caryophyllene oxide (4.52%), farnesol (1.35%), and b-carophyllene (3.09%).

Lippia species display enormous ecotype diversity in essential oil hydrodistilled from their shoots.

Litsea

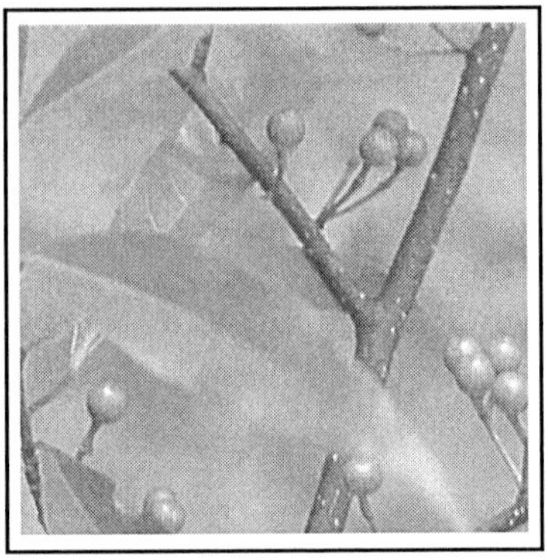

Scientific name: *Litsea cubeba* Pers.
Syn: *L. citrata*
Family: Lauraceae
Vernacular names: *Chineese*: May Chang; *Assamese*: Mezankari, Suntero, Taier, Tin-morcha, Dieng-se-ing

Litsea is a tropical tree with fragrant flowers and leaves, which emit a lemony odor when crushed. Oil is obtained from small pepper shaped fruits which resemble peppers. Steam distillation of fruits gives an essential oil with high citral content. It is used as a raw material source for isolation of citral. Litsea essential oil is rich in about 70% citral and has an intensely lemon-like, fresh, sweet odour. Due to its fragrance, litsea competes to a limited extent with lemongrass, another citral-rich oil. Litsea essential oil is used as a fragrance component in air fresheners, soaps, deodorants, colognes, toiletries and perfumes. It is also used to flavour fruit products. However, stem-bark of plant is used for treatment of eczema and scabies. Litsea fruits are used in north eastern region of India as pickle and condiment in meat, vegetables and curries.This plant, in fact, was an industrial crop of Assam during the period of Ahom administration. Fruits of litsea are edible, carminative and used as preservative for fish in Indonesia. Its fruits are used for treatment of headache, dizziness, hysteria and loss of memory in Indochina. In Indonesia, fruits are also used as substitute of cubeb pepper (*Piper cubeba*). Almost all parts of the plants are aromatic and yield volatile oils. A special kind of quality silk called "Mezankari Silk"which was once very popular was obtained by rearing "Muga-silk worm"(*Antheraea assama*) on its leaves.

Habitat and distribution

Litsea is a fast growing small to medium sized deciduous aromatic tree found in eastern Himalayas, Assam, Manipur and Meghalaya up to an altitude of 2700 m. Genus litsea has about 200 species distributed chiefly in tropical and subtropical countries of Asia and Australia. 65 species of litsea were described in the Flora of British India while 23 species were described from north east India. Litsea is native to China, Indonesia and some other parts of Southeast Asia, where it occurs mainly in mountainous regions. It is cultivated in central and eastern China south of Yangtze River. In Indonesia the species grows wild in Java, Sumatra and Kalimantan from 700 to 2300 m above sea level. China is the primary producer of *L. cubeba.* Chinese exports are of whole oil. Countries with capacity to fractionate essential oils and chemically convert isolates into products for flavour and fragrance industries are major importers of *Litsea cubeba* oil, i.e. the United States, countries of Western Europe and Japan. Total imports are probably of the order of several hundred tonnes annually, although trade in some years is estimated at 500 tonnes. A small quantity of oil is produced in Java and Indonesia. Citral content is the most important indicator of oil quality and an international (ISO) standard specifies a minimum value of 74%. Both Chinese *L. cubeba* oil and Indian lemongrass oil, with which it competes, are described in trade terms as "75 per cent", i.e. containing 75% citral.

The plant

This is a small tropical tree reaching a height up to 5-12 m with a stem diameter of 6-20 cm with lemon scented leaves and flowers. Stem bark of litsea is greenish, aromatic and warty. Leaves are aromatic, equilateral, narrowly elliptic to ovate-lanceolate, caudate-accuminate and thinly membraneous. Upper surface of leaf is bright green and lower surface is glaucous green. Petiole is slender, 2-4 cm long and purplish green. Flowers are umbels or corymbs peduncled in cluster of racemes. Dioecious female plant bears male flowers with rudimentary gynoecium while in female, flowers and stamens are staminode or less developed. Fruit berry type which resembles peppers, globose with succulent aromatic pericarp/pulp, black when ripen. In recent years, plant has become an important commercial crop for two kinds of value added essential oils; one as better substitute of lemon grass oil and the other as an alternative of *Eucalyptus citriodora* leaf oil. Despite several virtues and demand, adequate attention for commercial exploitation has not so far been paid on this important species.

Cultivation

Seeds of depulped fruits start germination at shorter period (4th week onwards) in comparison to seeds of normal pulped one (8th week onwards). Luke warm (30°-35°C for 30 min.) and cold water treatments showed higher germination (91.33% and 59.33%), respectively than normal seeds (36.67%). No germination was recorded for seeds of four weeks drying and storage under room condition. Litsea seeds when sown fresh germinate even after 44 weeks, although optimum germination was recorded (91.33%) in depulped seeds with luke warm water treatment. Seeds have a range of dormancy period from 28-308 days.

Plant population of Litsea in North East India has been decreasing considerably in its natural habitats. It is because seeds mature generally during July-August. Meanwhile, availability of soil water becomes scarce in its natural habitats. Ripe and mature fruits when detached naturally from tree fail to germinate due to unsuitable conditions and gradual drying. Seeds seen in natural habitats are ones expelled as bird excreta are depulped and hence able to germinate under favourable soil conditions.

Extraction and utilisation

Essential oil is extracted from the small pepper shaped fruits. Volatile oil is obtained by steam distillation. α-cis-ocimene (25.11%), 3, 7-dimethyl-1, 6-octadien-3-ol (16.85%) and n-transnerolidol (13.89%) are principal chemical constituents present in essential oil extracted from leaves of litsea. Steam distillation of fruits gives an essential oil with high citral content. Oil is a pale yellow mobile liquid with an intense, lemony, fresh-fruity odor sweeter than lemongrass but less tenacious. Volatile oil has specific gravity (26°C) 0.866-0.89, optical rotation about 7.0° and refractive index 1.46-1.48.

On GC analysis, five components representing 97.05% of oil are identified. Citral (A+B =51.90% + 40.68%) was detected as predominant (92.58%) component in oil. Other components (above 1%) concentration in oil are methyl-heptinone (1.46%) and linalool (2.84%). Occurrence of citral is major component (75.30%) in bark essential oil (0.6%) of litsea grown in north east India. The fruit essential oil grown in other countries contains 80% citral only. Hence, essential oil of plant native to North East India is considered superior.

Marigold

Scientific name: *Tagetes* spp.
Family: Asteraceae (Compositae)
Vernacular names: *Sanskrit*: Sandu, ganduga; *Hindi*: Guitera, genda; *Bengali*: Gendu; *Malayalam*: Chendumalli; *Tamil*: Tulukka, Samandi; *Kannada*: Seemeshamantige, chandumallige; *Telegu*: Bantichettu; *Gujarati*: Makhanala; *Marathi*: Rajia ka phul, Zendu; *Punjabi*: Mentok, Genda, Tangla; *Oriya*: Gendu

Marigold is a popular annual flower crop on account of its easy cultivation, wide adaptability, profuse flowering, short duration, wide spectrum of colours, shape, size and good keeping quality. There are African, French, Sweet scented, Stinking roger and stripped marigolds. African marigold flower oil is a fly repellent and is suggested as a modifier in hair lotions of bay-sum type. French type finds use in perfumery particularly in certain types of herbaceous fragrances like fougere and lavender. Oil is also employed in aldehyde oriented tabac bases. Oil of Stinking Roger finds applications in germicidal and microbicidal preparations due to presence of tagetone, which is toxic.

Planting of marigold is highly effective in controlling nematode population in soil. Intercropping marigold (*T. erecta*) with aubergine (*Solanum melongena*) lowers oot knot index than aubergine monocropping and growing aubergine and marigold alternately within rows give maximum check in *Meloidogyne incognita* population. Petals of marigold flowers (*T. erecta*) are commonly used as a feed supplement by poultry industry to enhance broiler skin and yolk colouration.

Habitat and distribution

African marigold, though its name suggests its origin from Africa, had its origin in Mexico. Sweet scented and stripped types are native of Mexico. Stinking Roger originated in South America, naturalized in north-west Himalayas between altitudes of 1250 and 2500 m above MSL like Simla in Himachal Pradesh. French marigold is widely grown in France and India up to an altitude of 1350 m. Marigolds are also cultivated in Australia and Kenya.

The plant

There are many species under the genus *Tagetes*.

T. erecta L. (African or Aztec Marigold): It is a major source of carotenoids, also grown as a cut flower and a garden flower grown for its medicinal values. They are erect and tall growing (up to 90 cm) plants having large globular flowers of diameter 15 cm and above, in shades of lemon yellow, bright yellow, golden yellow, orange and near white. Common cultivated varieties are Guinea Gold, Apricot, Primrose, Sun Giant, Fiesta, Golden yellow, Glitters, Happiness, Hawai, Crown of Gold, Honeycomb and Cupid, Pusa Narangi Gaintha and Pusa Basanthi Gaintha.

T. patula L. (French Marigold): These are bushy plants, 30-45 cm tall which flower profusely in singles or doubles. Flowers are in colour range of deep scarlet, red, primrose, yellow, golden yellow, orange and their combinations. Important cultivars are Rusty Red, Flame, Spry, Naughty, Marietta, Star of India and Harmony.

T. lucida Cav. (Sweet Scented Marigold): This is a perennial plant 30-40 cm tall with oblong-lanceolate, serrulate leaves. Flower heads are orange yellow and are borne in dense terminal clusters. They are often planted in gardens as a border plant.

T. minuta L. Syn: *T. glandulifera* Schrank. (Stinking roger): It is a highly aromatic annual 1-2 m tall; leaves 7-15 cm long, pinnatisect; leaflets 11-19, 4 cm long linear or lanceolate; flower heads pale yellow in corymbose clusters and black achenes.

T. tenuifolia Cav. Syn: *T. signata* Bartl. (Stripped Marigold): It is a branching annual, 30-60 cm high with pinnately compound serrate leaves and sweet smelling, bright yellow flowers.

Interspecific hybrids between African and French marigolds are also developed in USA. They are 50-60 cm tall with double flowers of 5-7 cm. Nugget, Show Boat, Red Seven Star, Red and Gold Hybrid are such hybrid varieties.

Cultivation

African and French marigolds are more widely cultivated as compared to other species. Marigolds in general, require a mild climate between elevations of 700-1500 m above MSL. They come up well on well drained rich loam or sandy loam soils with pH 5.6 to 6.5. They are propagated by seeds and cuttings; former is preferred for establishing tall, vigorous and heavy yielding plantations. Seeds are sown in nursery beds of 1.2 m width and 10-20 cm height and of 6 m length during May-June and transplanted within one month.

For African marigold, June planting was the best for maximum flower and seed yield in South Indian conditions. Irrespective of planting time (June,

July or August), the maximum flower production is generally achieved during September and October months since short day lengths during period are more congenial for flower bud induction and flowering in African marigold.

Main field should be ploughed well, incorporating 20 tonnes/ha of well decomposed FYM or compost. One month old seedlings are transplanted at 45x45 cm in African marigold and 30x30 cm in French marigold. A basal fertilizer dose of 112.5 kg N, 60 kg P_2O_5 and 60 kg K_2O/ha is applied for higher yield. Regular irrigation, weeding and hoeing are required to obtain large flowers. An additional dose of 112.5 kg N/ha at time of pinching with earthing up is required. Pinching consists of removing terminal portion of plant 30-45 days after transplanting. Initial flower buds are debudded to obtain bushy and compact growth.

Flower heads are harvested when they have attained full size. Flowers will be ready for picking in about 2½ months after transplanting and continue to bear flowers for another 2½ months. Flowers are harvested in evening along with a portion of stalks. Regular plucking of flowers increases flower production. Flower yield is 8-12 tonnes/ha for French marigold and 11-18 t/ha for African marigold.

Pest and disease management

Flower beetles, leaf hoppers, stalk borers and mites attack crop occasionally.

In poorly drained soils, *Phytophthora* foot rot occursand drenching with copper oxychloride checks it. *Sclerotinia sclerotiorum* causes stem rot which can be controlled by use of appropriate fungicides.

Extraction and utilisation

All parts of plant contain essential oil in varying concentrations. Oil is commercially obtained by steam distillation for 3-4 hours, absorbing distillate in petroleum ether or benzene. Prolonged distillation spoils fragrance. Marigold yields 0.02-0.08% oil with a turn over of 8-15 kg oil/ha/year.

Physico-chemical properties of oil

African marigold flower oil is reddish yellow in colour, possessing characteristic marigold odour and polymerizing readily in air. It has specific gravity (30°C) 0.936, refractive index (30°C) 1.5025, optical rotation 1.2, acid value 5.4 and ester value 33.5. Oil contains approximately ocimene 8.5%, limonene 14.03%, linalool 21.14%, linalyl acetate 13.75% and tagetone 40.38%. Leaf and stem oils are greenish yellow in colour.

French marigold oil is amber coloured, having a characteristic powerful fruity top note, reminiscent of green apples, with specific gravity (30°C) 0.8917, refractive index 1.492, optical rotation -3° 4□, acid value 6.1 and ester value 25. Tagetone 40.4%, linalool 22.1%, limonene 14.0%, linalyl acetate 13.8% and ocimene 8.5% are major constituents.

Oil of *T. lucida* is greenish yellow with specific gravity (15°C) 1.5218, acid value 6.0 and ester value 22. Major chemical constituent is estragol.

Rose

Scientific name: *Rosa damascena* Mill
Family: Rosaceae
Vernacular names: *Sanskrit*: Satapatri; *Hindi*: Fasli, Gulab; *Bengali*: Golap phul; *Malayalam*: Rosapoo, Paninirppu; *Tamil*: Gollappu, Rajappu; *Kannada*: Gulbihureu; *Telegu*: Rajappu; *Gujarati*: Gulabnu phul; *Marathi*: Gulab

Scented flowers are valued for worship and for making garlands. In addition, a variety of products such as rose oil, rose water, rose attar and rose otto are prepared from flowers. Rose oil called 'attar of rose' is one of the oldest and most valuable perfumery raw materials. Rose oil imparts characteristic fragrant top notes to perfumes and absolute adds lasting notes. A mixture of distilled oil and absolute combines advantages of both products. Distilled oil is employed in cases where solubility in dilute alcohol is important. Absolute is soluble only in high-proof alcohol and therefore used only in handkerchief perfumes or in cosmetics where solubility plays no role. In creams and powders, concrete is used. Bulgarian rose oil is used for flavouring certain types of tobacco and in a number of fruit flavours. Limited quantities of otto are employed in flavouring soft drinks and alcoholic liquors.

Rose has expectorant, laxative, sedative, aphrodisiac, antiseptic and anti inflammatory properties. Rose oil is used in aromatherapy as a mild sedative, antidepressant and antiinflammatory remedy. Rose oil has anti- microbial activity against *Staphylococcus aureus*, *Escherichia coli* and *Candida albicans*. Rose water is used from ancient times in medicinal preparations and syrups. Rose water makes a valuable lotion for inflamed and sore eyes. Rose is used as one of the ingredients in herbal eye drop preparation used in Ayurveda system of medicine, reported effective in a variety of infective, inflammatory and degenerative ophthalmic disorders. At marriages and other social functions, rose water is sprinkled on guests. Rose jam of Unani medicine is used as a mild laxative and tonic. With availability of cheap or synthetic substitutes like geraniol, use of rose flower for perfumery purposes has declined.

Habitat and distribution

Rose is indigenous to Europe. It is widely distributed in Europe and Middle East countries especially Iran, Afghanistan and Turkey. It is grown in Bulgaria, Russia, Egypt, France, India and Morocco. Main rose oil producers in the world are Turkey and Bulgaria and they obtain rose oil almost exclusively from *R. damascena*. They account for about 50% of world production of rose oil. Annual world production is around 15-20 tonnes. *Rosa damascena* is grown extensively in Uttar Pradesh and it is the major source of rose oil in India. Middle East countries are major consumers of rose oil based products.

The plant

Rose is a perennial erect shrub. Out of about 120 species of roses, three species are commercially used for production of rose oil, viz, *R. damascena* Mill, *R. gallica* L. and *R. centifolia* Linn. Out of these, *R. damascena* is the most important rose species for rose oil production from which most of high grade rose oil is produced.

Rosa damascena known as Damask rose or Bulgarian rose is an erect or climbing shrub with a productive life span of 10-15 years. It grows up to 3 m height; stems thorny; leaves pinnate; leaflets serrate. Flowers are terminal, solitary or corymbose, medium sized, pink; bracts rarely persistent; calyx tube persistent, globose, ovoid; lobes leafy, imbricate in bud; petals many, large; stamens many, inserted on disc; carpels many, rarely few, in bottom of calyx tube; styles subterminal, free or connate above; stigma thickened; ovule one and pendulous. Achenes cariaceous or bony, enclosed in fleshy calyx tube. Its petals and stamens contain high quality essential oil in higher quantities and is the most important species commercially grown for use in perfumery.

Rosa centifolia Linn. bears pink flowers and is grown for ornamental purpose and also for essential oil extraction. It is grown on a commercial scale in France. *R. gallica* is a deciduous shrub growing to about 1.5 m height, has sharp thorns, serrated leaves with 2-3 pairs of leaflets, deep pink or red flowers and scarlet hips. It is native to Middle East. *R. bourboniana* Desp is popularly known as Edward rose. It is cultivated mainly in south India. It is mainly used in making garlands, and for making rose water. *R. moschata* Herrm. is a profuse climber, bears with white single flowers in terminal clusters which possess musk like odour. *R. foetida* Herrm. is an erect bush of medium height with a few straight spines. Flowers emit a rich and heavy odour and are borne singly or in few-flowered groups. *R. gigantia* Collet is a rampant climber with thick, hooked prickles, producing fragrant white or pale-yellow flowers borne usually singly. *R. leschenaultiana* Wight & Arn. is a profuse climber with purplish branches bearing small hooked prickles. Flowers are large, pink, fragrant, borne in small clusters.

Cultivation

Rosa damascena is cultivated in temperate and subtropical climate. Performance in terms of flower and oil yield is much better in temperate climate having low rainfall. Mild and humid weather ideal at flowering. Plant grows the best in light and well drained, humus rich soils, but can withstand heavy soil. Acidic soils inhibit growth and reduce flower yield whereas alkaline soils with pH range of 7-9 are quite suitable.

CIMAP, Lucknow developed an improved variety of *R. damascena*, Noorjahan with higher oil content for commercial cultivation.

It is propagated either by cutting or budding. Rooted cuttings are usually used for field planting. Cuttings of 20-30 cm length and about 1 cm diameter taken from 10-12 month old shoots in January are first planted in a nursery or in poly bags. Dipping cut ends in IBA (200 ppm) before planting induces profuse rooting. Rooted cuttings of about 9-12 months age are planted in October- November, 1.5 m apart in pits filled with well rotten FYM @ 8-10 tonnes/ha. In temperate regions, a spacing of 2.5x2.5 m is given and a good crop requires 180 kg N, 100 kg P_2O_5 and 60 kg/hectare in 3-4 split doses annually. In subtropics, 90 kg N, 60 kg P_2O_5 and 60 kg K_2O is applied after pruning of crop in winter. Spraying NAA @ 50ppm and micronutrients @ 1% enhance flower yield.

Nitrogen application in 3 or 4 split doses increases secondary shoots, reduces dieback infection, induces early flowering and improves longevity of flowers. Nitrogen application positively influences bud circumference, flower diameter, petal length, petal breadth and flower yield. Rose plants need frequent irrigation during period of vegetative growth, flowering and just after pruning. In a year, 10-12 irrigations may be required. By third year onwards, flower yield will be economical.

Pruning, once or twice a year, preferably in October-February is essential for getting higher yield of flowers. First pruning is done at two years of age. In temperate climate, light pruning is given in autumn. Plants should be pruned up to a height of 50 cm from ground level. It takes 75-90 days for flowering after pruning. In subtropics, heavy pruning to a height of 15 cm from ground is given. Young sprouts, dead and diseased shoots should be removed twice in a year. Weeding and hoeing should be done after pruning. Generally, three weedings and hoeings are required annually. Annual weeds can be controlled by herbicides like simazine or atrazine applied at 3 kg/ha in light soils and 5 kg/ha in medium or heavy soils.

Rose flowers during March-April in plains and May-June on hills. In North India, there are two flowering flushes, main during April–May and the 2nd flush which constitutes 1/10 of main yield is during September-December. Peak flowering period is for 45 days. Sporadic flowering continues throughout year. Flowers are harvested from 5 am to 9 am in early morning, when they begin to open. Average yield of flowers is 2000-3000 kg/ha/year.

Pest and disease management

Pests

Aphids: (Macrosiphum rosae)

Aphids attack rose during flowering period. They can be controlled effectively by prophylactic sprays at 15 days interval with methyl demeton or metasytox (0.1%) or dimethoate (0.25%) or phosphamidon (0.1%).

Leaf caterpillars

Caterpillers of *Opherophtera frumanta, Malacosoma neustria, Orgyia antiqua* and *Archips podana* attack foliage during summer and rainy seasons. Caterpillars can be controlled by spraying endosulphan (0.2%) or quinalphos (0.15%) or fenthion (0.12%).

Red spider mites: (Tetranychus spp.)

Heavy infestation results in leaves becoming bronzed and falling prematurely. Wettable sulphur (0.4%) can be sprayed to control mites.

Rose thrips: (Thrips fuscipennis)

Thrips attack is a serious problem in rose during flowering period. Flower buds get discoloured and crinkled and fail to open causing significant reduction in flower yield. Sprays of mercaptothion EC (0.2%) or fenitrothion EC (0.1%) at an interval of 10 days controls thrips.

Caspid bugs: (Lygocoris pubulinus)

Pale green bugs suck sap from shoot tips and young flower buds, causing flower petals to develop small holes. Prophylactic application of carbofuran 3G or phorate 10G @ 10 kg/ha in soil after pruning and at bud formation keeps damage under control.

Rose scale: (Aulacaspis rosae)

Scales occur on stems and suck sap from stems, resulting in drying of shoot. Foliar sprays of monocrotophos (0.15%) or carbaryl (0.3%) or metasystox (0.15%) are recommended.

Termites

Different species of termites feed on roots of rose. Termite attack can be controlled by drenching chlorpyriphos EC 5ml/l.

Diseases

Black spot: (Diplocarpon rosae)

Fungus develops black spots on leaves resulting in leaf fall. Sprays with captan (0.2%) or benomil or carbendazim (0.1%) is recommended for control of this disease.

Rust: (*Phrogmidium mucronatum*)

Orange coloured pustules develop on leaves and at later stages become dark brown. Spraying fungicides like plantavax (0.2%) or mancozeb (0.3%) are recommended for control.

Powdery mildew: *(Sphaerotheca pannosa)*

Small, white, powdery fungus appear on leaves, stems and occasionally on flowers. Dinocap (0.1-0.15%) or wettable sulphur (0.4%) should be sprayed at 15 days interval for control of disease.

Downy mildew: (*Peronospora sparsa*)

Small reddish purple areas develop on the youngest leaves resulting in leaf distortion. Repeated sprays of mancozeb (0.2%) or captafol (0.3%) or copper oxychloride (0.3%) should be given for effective control.

Rose rust: (*Phragmidium subcortium*)

Leaves, flowers and shoots are infected by rust and cause leaf and bud fall. Removal and destruction of fallen leaves and buds is very important in managing disease. Spraying of carbendazim @ 0.1% twice during flowering period (March-May) and benomyl @ 0.1% periodical spraying check the disease.

Virus diseases

Rose mosaic is the most common virus disease. Chemical control measures to kill various virus vectors should be followed to contain the problem. Infected plants should be up-rooted and destroyed.

Extraction and utilisation

Rose oil is extracted from flowers by distillation for 2-3 hours or by extraction with volatile solvents. Flowers can be stored in clean cold water for a period of 3 days without any loss in oil recovery or change in oil quality. Average oil yield is 0.03% on fresh weight basis (f.w.b). Water distillation is a popular method for obtaining rose water. In India, a major share of flowers produced are used for rose water distillation. It is then matured for a few months and then diluted and bottled. '*Rose attar*' is obtained by collecting water distillate over sandal wood oil. '*Otto of rose*' is prepared by water distillation of rose flowers and redistilling distillate 2-3 times till it gets saturated with oil dissolved in it. Then it is chilled and oil drops floating on surface of water are removed. Yield of oil varies between 0.01 to 0.04% based on method used and flowers used.

In India, rose water, rose attar, rose oil and gulkand are major products from Damask rose. Production of concrete and absolute, richer in volatile constituents are limited. A concrete contains fragrance principles (monoterpenes and sesquiterpenes), fatty acids, their methyl esters, paraffins and a few other high molecular weight compounds. Total extract is then dissolved in absolute alcohol, chilled and filtered to get rose absolute and turnover is approximately 54-64 % of concrete. Major components in both are phenyl ethyl alcohol, citronellol, nerol, geraniol and nonadecane.

Physico-chemical properties of oil

Rose oil is a colourless liquid, but on aging develops an amber colour. It has specific gravity (30°C) 0.8845, refractive index (30°C) 1.4657, optical rotation 2.5°, Acid value 2-6, ester value 18.7 and total alcohol content 81.6%. Essential oil contains various alcohols, aldehydes, ketones, esters, phenols, terpenes and acids. Major components are citronellol 38%, paraffins 16%; geraniol 14%; nerol 7%; β-phenyl ethanol 3%; eugenol methyl ester 3%; linalool 2%; ethanol 2% and farnesol 1%.

Rosemary

Scientific name: *Rosmarinus officinalis* Linn.
Family: Lamiaceae (Labiatae)
Vernacular names: *English*: Rosemary, Romemary; *Hindi*: Rusmari; *Tamil*: Rosemary

Rosemary is a dense evergreen undershrub with lavender-like leaves and a characteristic aroma. Its essential oil is used almost wholly in perfumery industry in production of soaps, detergents, household sprays and other such products. It is an excellent fixative material. Oil contributes a strong, fresh odour, which blends well with various other oil odours and also serves to mask unpleasant smells of certain other ingredients. It is used in shampoo, toilet soaps and medicine. Rosemary oil is known to have antimicrobial activity. It can kill 90-100% of mosquito and larvae of *Culex quinquefasciatus*. It is carminative and mildly irritant. It is used in formulations of compounded oils for flavouring meat, sauces, condiment and other food products. It is used as a culinary herb. Rosemary is a popular spice in many Western countries, but its usage is most popular in Mediterranean countries, especially Italy and France. Rosemary does not lose its flavour by long cooking. Fresh leaves have a more pure fragrance and are therefore preferred whenever available. Flower water obtained on distillation is used as eyewash. Wood is used to make lutes and other musical instruments. Scent of rosemary is an effective memory stimulant. This might make a nice potted plant for your desk at work or where kids do their homework.

Habitat and distribution

Rosemary is an attractive evergreen shrub with pine needle-like leaves. It's trusses of blue flowers last through spring and summer in a warm, humid environment. Plant comes up well in Mediterranean climate. It is indigenous to South Europe, Asia Minor and North Africa. It grows wild on Mediterranean shores and in Spain, Portugal, Morocco and Algeria and is cultivated in Spain, Tunisia, Yugoslavia, France, Italy, North Africa and India. In India, it is grown in Nilgiris. Plant is susceptible to frost injury. In cooler areas, it can be cultivated in summer season. In a warm climate, plant remains in same location for up to 30 years, but in climates where freezing temperatures are expected, plants may be grown in pots so that it can be brought indoors in winter. Rosemary succeeds best in a light, rather dry soil. On a chalk soil it grows smaller, but is more fragrant. Silver and gold-striped types are not quite so hardy and hence require light dry soil, preferably lying over chalk. Neutral to alkaline pH is suitable.

The plant

It is an evergreen dense highly branched herb or undershrub growing upto 1m in height. Botanical name Rosmarinus is derived from old Latin for 'dew of the sea', a reference to its pale blue dew-like flowers and the fact that it is often grown near sea. Evergreen leaves of this shrubby herb are about 2.5cm long, linear, revolute, dark green above and paler and glandular beneath, with an odour pungently aromatic and somewhat camphoraceous. Flowers are small and pale blue. Much of active volatile principle resides in their calyces. There are silver and gold-striped varieties, but green-leaved variety is the kind used medicinally. Tradition says that rosemary grows for thirty-three years, until it reaches height of Christ when he was crucified, then it will die. Long slender branches bear many sessile opposite leaves, smooth and green, woolly whitish and glandular beneath, 2-4 cm long, almost cylindrical and folded inwards; flowers situated in small clusters towards ends of branches; calyx is 2-lipped with an upper single broad oval lobe and a lower two segmented triangular lobe; corolla is two lipped with two violet stamens and a long style projecting from it; fruit is an oval 4-sectioned cremocarp.

Cultivation

Plant is propagated through seeds or from cuttings of twisted wood of non-flowering branches in early summer or layer established branches. Propagation by cuttings is more common. Cuttings should be 15 cm long and leaves removed from basal half portion. They are put in nursery beds of sandy soil at a depth of about 10 cm. Layering may be readily accomplished in summer by pegging some of lower branches under a little sandy soil. Finest plants are said to be raised from seed. Main field is prepared well incorporating 10-15 tonnes/ha of organic manures. The 50 days old rooted cuttings are transplanted in rows, 120cm apart with a plant to plant spacing of 30-40 cm. Fertilizers are applied at 100:40:40 N, P_2O_5 and K_2O/ha, N being applied in 4-5 split doses during each year. Irrigation is needed when soil is depleted of water during non-rainy period. Field of rosemary is kept weed free by regular weeding and hoeing. Intercultivation keep soil loose and clean from weeds and promotes proper plant growth and development. First harvest is done after 180 days of planting and subsequent harvests at 120 days interval after observing 50% flowering; thus two harvests are taken during first year and three harvests in subsequent years. Crop is harvested at 20 cm above ground level. Shoots are cut when they have reached their maximum size but before they have become woody. Hard wood imparts an undesirable turpentine odour to essential oil. Frequent cutting of bushes after 2-3 years keeps them free from becoming leggy and promotes formation of numerous young shoots. A well maintained rosemary crop gives economic yield for 4-5 years or even more. Average herbage yield per annum is 30 tonnes/ha.

Extraction and utilisation

Freshly harvested twigs and leaves are steam distilled to obtain essential oil of rosemary. Steam distillation at 2-3 times atmospheric pressure gives an oil yield of 1.0-1.5% of freshly harvested plants and 1.5-2.5% of dried leaves. Plant contains some tannic acid together with a resin and a bitter principle and a volatile oil. Chief constituents of oil are borneol, bornyl acetate and other esters, a special camphor similar to that possessed by myrtle, cineol, pinene and camphene. It is colourless, with odour of rosemary and a warm camphoraceous taste. Chief adulterants of oil of rosemary are oil of turpentine and petroleum. Rosemary yields its virtues partly to water and entirely to rectified spirits of wine. Total oil yield per hactre is on an average 400 l/ annum.

Physico-chemical properties of oil

Essential oil of rosemary has following properties.

Specific gravity at 15°C	0.894-0.913
Refractive index at 20° C	1.466-1.468
Optical rotation	0° 43' to 13° 10'
Esters as bornyl acetate	1.8-7.0%
Total alcohol as borneol	8.4-14.3%

Essential oil contains chemical components as α-pinene 7-24%, camphene 3-9%, β-pinene 0.6-1.5%, sabinene trace, myrcene 2-19%, α-phellandrene 0.1-1.8%, α-terpinene 0.5-1.5%, limonene 3.3-5.4%, 1,8-cineole 15-20%, γ-terpinene 0.2-1.8%, p-cymene 2-3.7%, terpinolene 0.1-0.6%, camphor 0.1-0.6%, copaene 0.1-0.3%, linalool 14-17%, terpinen-4-ol 0.9-1.8%, caryophyllene 0.9-2.9%, α-terpineol 1.8-3%, thymol 0.1-0.7% and carvacrol 0.1-2%.

Skimmia

Scientific name: *Skimmia laureola* Sieb. & Zucc. ex. Walp Syn: *S. arborescens* T. anders ex. Gamble
Family: Rutaceae
Vernacular names: *Hindi*: Kasturchara, Kedarpatta, Nora; *Punjabi*: Barru, Ner, Shalangli, Patrang; *Kasmiri*: Patar, Ner, Nera

Skimmia oil obtained from leaves is used in high grade perfumes. It is a rich source of linalool and linalyl acetate. Principal constituent of essential oil, linalyl acetate is of much importance in manufacture of soaps, cosmetics, perfumery and flavouring. Oil resembles French petitgrain bigarade oil in physico chemical properties and hence being used as its substitute. Oil is antiseptic, effective against *Staphylococcus* and *Streptococcus*. In China, it is used for darkening hair and also for hair washing.

Leaves are used as incense in various religious Hindu rites, eaten in curries by hill tribes and also used for flavoring food in Kashmir. Leaves are also used as an incense to purify air to protect against small pox. Seeds yield around 27% fixed oil and red seed coating yields pelargonidin. Skimmia is valued as an ornamental tree, ideal for border planting in gardens because of their handsome foliage and red berries. Hard and heavy wood is used for making handles for agricultural implements.

Habitat and distribution

Skimmia grows copiusly as undergrowth in moist shady areas in mixed coniferous forests of India. It comes up well in high rainfall areas. Plants grow abundantly at higher elevations in western Himalayas extending to e Asia and Philippine Islands. It grows in Jammu & Kashmir, Himachal Pradesh and Utter Pradesh at altitude between 1800 to 3200 meters.

The plant

It is a slow growing, small, erect, evergreen, glabrous perennial shrub attaining a height up to 3 m. Whole plant is aromatic. Leaves are simple, thick, leathery, alternate, lanceolate or oblong, 7.5–15x2-3.8 cm, entire, with prominent midrib and without conspicuous secondary nerves. Yellowish green or yellowish white, sweet scented, small flowers are arranged in erect, compact, terminal panicles. Calyx 4-5 lobed and imbricate. Petals 4-5, valvate or slightly imbricate. Stamens 4-5, imperfect in female flower. Ovary entire, 2-5 celled; style single or absent, stigma capitate, 2-5 lobed; ovules solitary. Fruit is a fleshy drupe, ovoid, 1-1.5 cm long, scarlet red, 1-3 seeded with cartilaginous kernels.

Cultivation

Plant is usually collected from wild for extraction of oil. Skimmia thrives best in a moist loamy soil. Propagation of plants can be done by seeds, stem cuttings and root suckers. After harvest, stems after removal of leaves for oil extraction can be used as planting material. Rooting occurs within one month.

It is an evergreen shrub, but defoliation occurs during snowfall during winter December-February. New flushes appear in March-April and leaves mature enough in May-June for first harvest. Collection period spreads well over six months, from May to October, during which two crops of leaves can be harvested, in June and October. A seasonal variation in concentration of linalyl acetate and linalool is reported. Oil content declines gradually after August and concentration of linalyl acetate is maximum in May and August-September.

Extraction and utilisation

Leaves are either distilled fresh or air dried to lose up to 60% of moisture especially when long distance transport is done before distillation. Proper air drying is a problem in high elevations due to unfavourable climatic conditions prevailing in such areas. Storage of fresh leaves beyond 3-4 days results in fermentation and thus in deterioration of quality of oil. Steam distillation will take about 6 hrs. Fresh leaves with young twigs yield 0.5-0.8% clear bluish oil whereas air dried material yields 1.4-3.0% oil. Linalyl acetate content varies from 39-70% depending up on time of collection of leaves and distillation method employed. Stem contains on an average 0.1% oil. Essential oil extraction is generally done by steam distillation.

Physico-chemical properties of oil

Oil characteristics are reminiscent of petitgrain.

Specific gravity	0.8875-0.9220
Optical rotation	- 40° 45′
Refractive index	1.4647-1.4757
Acid value	0-3.7
Saponification value	81.1-197.9
Ester content	28.4-69.3
Linalool content	10.4-18.8
Solubility in alcohol	1 in 2-3 parts

Linalyl acetate is the major constituent of essential oil contributing to 60-71.5%. In addition it contains linalool 17.5%, α-pinene 0.15%, β-pinene 0.16%, myrcene 0.4%, β- phellandrene 1.2%, geraniol 0.67%, nerol 0.27%, citronellyl formate 0.65%, citronellyl iso-butyrate 0.14%, methyl heptanol 0.34%, and citral 0.1%, *p*-cymene and γ-terpinene 0.24%, camphor 0.50%, citronellal 0.75%, borneol 0.05%, eugenol 1.68%, caryophyllene 0.44% and humulene and farnesene in traces. In addition to the two major constituents, linalool and linalyl acetate, two non terpenoids, pregeijererene and its isomer geijerene are also reported. Of these two C_{12} hydrocarbons, pregeijerene contributes to typical sweet and pleasant aroma of *S. laureola* oil. Besides essential oil, leaves and bark contain a quinoline alkaloid, skimmianine and coumarins namely umbelliferone, bergapten and isopinellin. Seeds yield around 27% fixed oil.

Sweet Flag

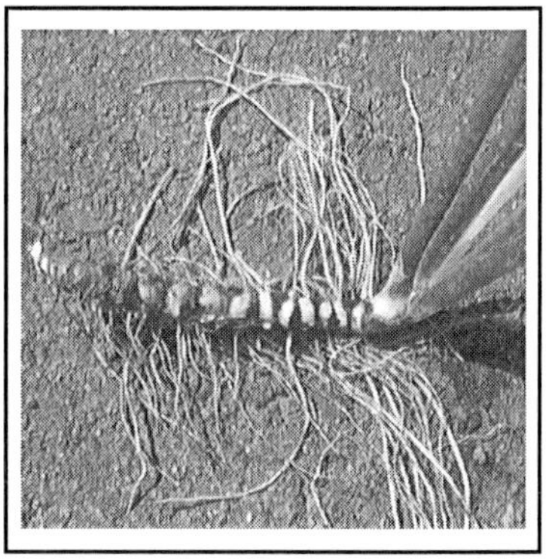

Scientific name: *Acorus calamus* Linn.
Family: Araceae
Vernacular names: *English*: Calamus; *Sanskrit*: Vacha, *Mangalaya*, Jatila, Jalaja; *Hindi*: Vacha, Ghoorvacha; *Kannada*: Vajaey, Bajegida; *Bengali*: Bacha, *Arabic*: Vash; *Tamil*: Vashambu; *Malayalam*: Vayampu; *Kashmiri*: Vabi

Sweet flag, *A. calamus* is a popular medicinal plant used in several ayurvedic formulations in India. It is an important medhya drug, capable of improving memory and intellect. It is used for treatment of cough, bronchitis, odontalgia, inflammations, gout, epilepsy, convulsions, depression and other mental disorders, tumors, dysentery, skin diseases, numbness and general debility. Rhizome is an ingredient of preparations like 'vachaditaila', 'ayaskrti', 'kompancadi gulika' and 'valiya rasanadi kashaya'.

Habitat and distribution

Sweet Flag, native of Europe is a perennial plant growing wild on edges of swamps, on banks of rivers and ponds in North America, Europe and Asia. Commercial producers of sweet flag are Russia, Central Europe, Romania, India and Japan. It is distributed throughout tropics and subtropics, especially in India and Sri Lanka.

This plant is available both under wild and cultivated conditions in India. It is plentiful in marshy tracts of Kashmir and Sirmoor in Manipur and Naga Hills. It is estimated that 77% of produce comes from a cultivated area of about 80 ha and rest 23% gathered from wild. In view of acute scarcity of this material, farmers in Karnataka are undertaking cultivation of this annual crop in Tumkur district. Total production of sweet flag in the country is estimated at 573 tonnes/annum.

The plant

Sweet Flag is a rhizomatous, branched, semi-aquatic, creeping herb, usually perennial and rarely biennial. Light brown to pinkish brown, rhizomes 1.2-2.5 cm thick are cylindrical, compressed and spongy. Leaves are bright green, measuring 0.9 to 1.8 m in length and 1.7-3.5 cm in width, distichous and ampexiant at base. Scape or flower stem arises from axils of outer leaves, which it much resembles, but is longer and solid and triangular. From one side, near middle of its length, projecting upwards at angle from stem, it sends out a solid, cylindrical, blunt spike or spadix, tapering at each end, from 6-12 cm in length, often some what curved and densely crowded with very small greenish-yellow flowers. Each tiny flower contains 6 stamens enclosed in a perianth with 6 divisions and surrounding a three celled, oblong ovary with a sessile stigma. Flowers are sweet scented and so formed that cross pollination is ensured.

Flowers are light brown, sessile, densely packed in cylindrical spadix; perianth polyphyllous; stamens are filaments, linear, membranous and flattened. Anthers reniform, ovary 2-3 chambered conical and superior. Fruits 3-celled, oblong, turbinate, berries with a pyramidal top. Seeds are a few and pendant from apex of cells.

Cultivation

Acorus is a hardy plant growing from tropical to subtropical climates. It needs a good and well-distributed rainfall throughout year. It needs ample sunlight during growth period as well as after harvest for drying rhizomes. It may be cultivated in any good but fairly moist soil. It is usually grown in areas where rice can be grown. It comes up well in clayey soils and light alluvial soils of river bank. Field is laid out and prepared exactly as for rice, irrigated sufficiently and after ploughing twice, watered heavily and again ploughed in puddle. Sprouted rhizome pieces are used for planting and pressed into mud to a depth of about 5 cm at a spacing of 30x30 cm. Rhizomes are planted in such a way that plants in second row come in between plants of first row and not opposite to them. FYM is applied at 25 tonnes/ha. Fertilisers are applied at 25:50:60 kg N, P_2O_5 and K_2O/ha/yr. Whole of FYM and 1/3 of N, P_2O_5 and K_2O are added in field during March-April as basal dose. Remaining 2/3 of nutrients is given in two equal split doses at 4 months and 8 months after planting. Field is regularly irrigated. About 5 cm of standing water is maintained in the field in beginning. Later, it is increased to 10 cm as plant grows. Field is regularly weeded. About 8 weedings are required. At each weeding, plants are pressed into soil. Plant is attacked by mealy bugs. Both shoot and root mealy bugs can be controlled by spraying shoot and drenching roots of grown up plants with 10 ml methyl parathion or 20ml quinalphos in 10 litres of water. Crop is ready for harvest at end of first year. Field is dried partially so that sufficient moisture is left in soil to facilitate deep digging. Leaves start turning yellow and dry, indicating maturity. Rhizomes grow to a depth of up to 60 cm and shall have a spread of about 30-60 cm. Harvesting is done at about 11-12 months during Febrauary-March by carefully collecting rhizomes after digging soil. Rhizomes are washed thoroughly to remove adhering soil, all fibrous roots removed and then cut into pieces of 5-7.5 cm in length. Rhizome yield is about 10 tonnes/ha.

Extraction and utilisation

Rhizomes, roots and leaves of sweet flag contain essential oil. Oil is usually distilled from rhizomes only. Dried rhizomes contains about 2-3% oil. Rhizomes are dried till they become brittle. Dried rhizomes can be utilized in several ways; they can be stored for some time or used as such for preparing medicines or steam distilled to collect essential oil. Volatile oil is yellow resembling that of patchouli. Rhizomes are rendered free of fibres when later becomes brittle after drying. Resulting thick root stock becomes richer in essential oil content. To have an optimum yield of essential oil, dried and unpeeled rhizomes are granulated immediately in a suitable disintegrator before their charging in to stills. This improves yield from 1.5-3.5% to 4.6%.

Physico-chemical properties of oil

Specific gravity (15 °C)	1.0694
Refractive index (20 °C)	1.5030
Optical rotation	6° 12'
Acid value	1.4
Saponification value	5.1
Saponification value after acetylation	16.6
Methoxy content	36.7%
Solubility in alcohol (8°C)	Soluble in 1.5 volumes of 70% alcohol

Important constituents of Indian oil are asarone 82%, calamenol 5%, calamine 4%, calamenone 1%, methyl eugenol 1%, eugenol 0.3%, alpha-pinene+camphene 0.2%, besides traces of palmitic acid, n-heptylic acid, asaroaldehyde and a butyric ester.

Sweet Marjoram

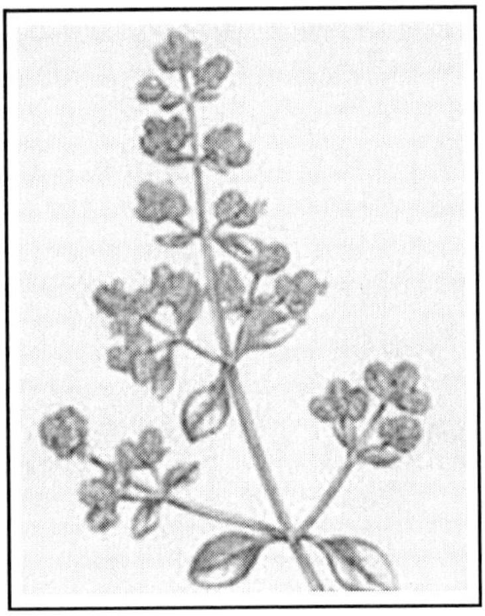

Scientific name: *Marjorana hortensis* Moench; Syn: *Origanum marjorana* Linn.
Family: Lamiaceae (Labiatae)
Vernacular names: *Hindi*: Murwa, Marawaa, Maruvam; *Sanskrit*: Marubaka, Sukhatmaka, Maruvaka; *Kannada*: Maruga; *Tamil*: Maruvu, Marukkolundan; *Bengali*: Murru; *Malayalam*: Maruvamu

Marjoram (*M. hortensis*) is an aromatic perennial herbaceous plant. Whole plant, leaves, seeds and essential oil are put to a variety of uses. Fresh springs of this plant are also used to flavour garlands, bouquets and for higher adornment. Traditionally, leaves of plant are used fresh or dried and are highly esteemed as a condiment for seasoning food. Fresh leaves are employed as garnish and incorporated in salads and vinegar. Dried flowering tops are used for sachets and potpourri. Aromatic seeds also find use in confectionery. Steam distillation of leaves and flower heads give a volatile oil known in trade as oil of sweet marjoram. Oil is mainly employed in flavouring food and is much in demand in food processing industries for flavouring canned food. To a small extent, it is used in high grade flavouring preparations and in perfumes and soaps. Medicinally, it is used as an external application for sprains, bruises, stiff and paralytic limbs, toothache and for hot fomentation in acute diarrhoea. Sweet Marjoram is carminative, expectorant and tonic; leaves and seeds are astringent.

Essential oil obtained from leaves of marjoram has antimicrobial and nematicidal activity.

Habitat and Distribution

Plant is native to southern Europe, North Africa and Asia minor. It is an important kitchen herb with a wide range of uses and is cultivated throughout Europe and Mediterranean region as well as North and South America. France and Egypt are the biggest exporters of marjoram oil in the world. In India, crop is cultivated only on a small scale. It is grown in home gardens in Karnataka, Andhra Pradesh and Tamilnadu. It is primarily a warm climate plant and as such rather sensitive to cold. It requires sunny position, though midday shade is preferred. Marjoram prefers well drained and nutrient rich soil for its successful growth. It has a strong flavour when grown in rich soil.

The plant

Marjoram is a perennial herbaceous plant, grown often as an annual, because it does not survive well in wet and cold winters. It grows to an upright, compact bush about 20-40 cm height, with a woody main stem and many softer branches. Leaves are oblong-ovate, soft, matte green with a marvelous, sweet, spicy, pleasant smell. Flowers are small, whitish or purplish, arranged in terminal clusters; seeds minute, oval and dark brown.

Cultivation

Crop is usually grown vegetatively through cuttings although it can also be grown by seeds, which is a little cumbersome method. Planting time is one of the important factors in obtaining optimum yields from a unit area. Seeds can be sown twice a year during June-July and September-October. They are sown in well prepared nursery beds in rows spaced at about 15cm apart. About 50 gm seeds give sufficient seedlings to plant a hectare area. As the seeds are tiny, they are mixed with dry sand (1:10) and broadcasted in rows of nicely prepared nursery beds. A thin layer of sand is spread uniformly over seeds. Seeds germinate after 8-10 days. Seedlings are ready for transplanting when they are two months old. Commercial propagation is done through cuttings. Stem cuttings obtained from second ratoon are the best for planting. Herbaceous cuttings of 15 cm length are taken from healthy plant. They require one month to root. Individual polythene bags of small size may also be used to raise plants through cuttings. In this method, roots do not get damaged while transplanting and establishment rate is 90-95%. Healthy seedlings or rooted cuttings with well developed root system are used for transplanting in main field at a spacing of 15x7.5 cm. First harvest is done after 90 days of planting, when crop starts flowering. While harvesting, about 10 cm of shoot above ground level is left for further growth. Subsequent harvest can be carried out after 45 days of first harvest. On an average, about 4-5 harvests are taken annually. Crop can be retained up to seven cuttings under good management.

Extraction and utilisation

To get oil of export quality, first harvest should be done only after 90 days of planting. Oil is extracted by steam distillation. Oil content in fresh leaves of marjoram ranges from 0.3-0.5%. Dry plant gives up to 1% oil. On an average 35-40 kg oil is obtained from one hactre of the crop per year.

Physico-chemical properties of oil

Major components of the oil are terpinen-4-ol (30.55%), (z)-sabinene hydrate (16.86%) γ-terpinene (6.37%), ρ-cymene (5.83%), (E)-sabinene (4.81%), α-terpineol (3.71%), (E)-sabinene hydrate (3.63%), γ-terpinene (2.67%) and linalyl acetate (2.41%)

Leaf oil has following physico-chemical properties.

Specific gravity at 33°C	0.845
Phenolic content	48.00
Refractive index at 14.5°C	1.5060
Optical rotation	40.30°
Acid value	5.60
Ester value	38.00
Ester value after acetylation	44.26

Gas chromatographic analysis showed that leaf oil consisted of α-pinene 6.32%, β-pinene 7.14%, di-limonene 2.94%, camphene 3.92%, bornyl acetate 3.51%, β-ocimene 2.94%, α-thujone 3.12%, 1,8 cineole 7.32%, myrcene 2.84%, ρ-cymene 14.0%, d-α terpineol 4.8%, methyl chavicol 3.20%, carvacrol 36.70%, geraniol 7.94%, eugenol 26.0%, azulene 3.92% and caryophyllene 7.60%.

Tea Tree

Scientific name : *Melaleuca alternifolia* (Maid. & Bet.) Cheel. Syn: *M. linariifolia* var *alternifolia* Maid and Bet.
Family: Myrtaceae

Tea tree oil has antibacterial, antifungal, antiviral and anti-inflammatory properties and is employed in commercial pharmaceutical market. It is a broad spectrum disinfectant and is the greatest natural antiseptic of world. Oil is applied externally on skin for wound healing, for treating boils, rashes, cuts, tooth aches, gum infections, mouth ulcers, toe nail infections, athlete's foot and scrapes and insect bites, stings and carbuncles. Oil is added in soaps and body washes. It is active against organisms which cause paronychia and aids in relief of pain, pressure and swelling.

Tea tree is a traditional aboriginal remedy. Leaves are crushed in hand and the volatile oil is inhaled or used as infusions for cough, cold, head aches, sore throat and skin infections. It is taken internally for chronic and acute infections and also used in mouth washes and gum disease. In home remedies, tea tree oil is added to a vaporizer to cure chest congestion, to shampoo to destroy head lice and to a bath to remove persistent body odor.

A gel containing 5% tea tree oil was effective in treatment of acne (*Propionibacterium acnes*). Oil showed efficacy against *Candida albicans* and *Trichomonas vaginalis*, that cause vaginal infections. A solution containing 0.5% tea tree oil offers protection against *Pityrosporum ovale*, a dandruff causing fungus and head lice. It is effective against *Staphyloccus aureus*. Cytotoxicity evaluation of tea tree oil supports its use only in topical application and not for ingestion purposes. It may be toxic when ingested internally in large doses or by children. Antimicrobial activity of tea tree oil and its constituents is in order of α– terpineol > terpinen-4-ol > tea tree oil > 1,8- cineole.

In aromatherapy, tea tree oil is used in recipes for colds, sore throat, coughs, acne, tooth aches, pyorrhea and sunburns. Mouth sores and gingivitis can be checked by mouth wash prepared by adding a few drops of tea tree oil into a cup of warm water. Oil gives protection against radiation damage. Fumigation of stored agricultural products is done using the oil against post harvest fungal pathogens. Leaves are used as a substitute for tea and also to flavour beer. Its foliage when steeped in water produces a brown tincture resembling tea. Tea is suitable for hedges and is popular as garden plants in Australia and other tropical areas.

Habitat and distribution

Plant is native to eastern Australia. Tree oil is used by Australian aborigines for several centuries for its medicinal properties. They also brewed the leaves into a spicy tea. Tree grows in inaccessible parts of dividing range of eastern Australia and is very sparsely distributed and confined to warmer (27-32°C) swamps by the coast. Occurrence of wild stands in swampy areas makes harvesting difficult. It is frost sensitive. It thrives in a relatively small area of New South Wales, in swampy low-lying land surrounding flood-prone river systems. Now its cultivation has spread all over world. USA and Europe are major users of the oil. More than 700 tonnes are produced annually.

The plant

Although there are over 240 species of trees in *Melaleuca* family, *M. alternifolia* produces quality tea tree oil suitable for therapeutic use. It is known as narrow leaved paper bark tree/ opposite-leaved bottlebrush/ snow in summer. It develops into a small, evergreen tree of 5-7 m height, usually with a single trunk with flaky, exfoliating bark and white spring and summer flowers. Tree growth is moderate. Branches may be many or a few, and branching occurs low or high. In dense stands, a few branches grow almost vertically. Roots at base of stem are partly exposed and sometimes adventitious roots are also found.

Bark consists of thin paper like layers of cork separated by thin fibrous layers. Bark which later peels off naturally is used by gardeners for filling hanging basket. Wood is termite and water resistant and used in building construction and also as firewood. Leaves are stiff, citrus scented with a uniform green color, glabrous, very narrow, alternate, ovate to lanceolate, entire and stouty petiolate.

Flowers are produced in dense clusters along stems, each flower with fine small petals and a tight bundle of stamens. Flowers are creamy white, sometimes yellow tinted, sessile in open or condensed terminal spikes. It is known as snow-in-summer in its native habitat owing to its white flowers in spring and summer. Flowers are bisexual. Fruit is a woody capsule, globular, less than 0.5mm diameter, occurring in elongated clusters. Seeds are minute, brown, normally 10,000 numbers per gram and with very low viability.

Plant is cross pollinated by insects and it hybridizes freely with other members of the genus. Genus *Melaleuca* is closely related to *Callistemon*. Stamens are generally free in *Callistemon* but grouped into bundles in *Melaleuca*.

Related species *M. cajupti* yields cajuput oil and *M. quinquenervia* (syn. *M. smithii, M. maidenii*) yields niaouli oil of commercial use. *M. viridiflora* (broad leaved paper bark tree) and *M. linariifolia, M. dissitiflora, M. bracteata* (black tea tree) also provide valuable essential oil similar to tea tree.

Cultivation

Tea tree grows well on a range of soils with adequate soil moisture but it can cope well with dry situations. It is not adapted to shady conditions and requires full sun for the best performance. It does well in poorly drained soils such as swampy areas, drainage lines and river banks. Plant prefers acid and neutral soils. It is not very cold hardy and does not tolerate temperatures below -7°c for long period.

Direct seed sowing is rarely practiced. Seeds are either sown in seed beds composed of coarse sand and loam or in pots and later transplanted. Seeds are sown thinly at the rate of 0.5 g/m^2. Seeds germinate readily. Seeds sown in seed trays are planted out singly in pots when 4-6 weeks old. Field planting is done after frost. Adequate soil moisture should be assured at time of transplanting. Field establishment is low when transplanted from nursery beds compared to potted plants. Cuttings are also used as planting material after rooting in pots. Plant population will be 30,000 to 35,000 per hectare.

During initial years of field establishment, seedling should be protected from cold. Fertilizer application is generally not practiced but occasional use of slow release fertilizers is beneficial. Crop requires low maintenance. In Australia, tea tree is attacked by larvae of hepialid moths *Aenetus ligniveren* which first burrow horizontally into trunk and then vertically down.

First harvesting is done at 15-18 months and subsequently at 12-15 month intervals. Lleaves and branches are harvested throughout the year and distilled. Stems should be cut when diameter is less than 2 cm and at 5-10 cm above soil level. Foliage yield will be 8-10 tonnes/ha. Leaf oil content is lower in cooler months. In cold regions, harvesting should be so planned that emergence of new shoots will not coincide with frost. Oil yield will be around 70 kg/ha during first year and increases and stabilizes by third year to around 150 kg/ ha.

Extraction and utilisation

On steam distillation, fresh leaves yield 2-3 % lemon tinted essential oil. Tea tree oil is non-irritant. Crude oil is dark yellowish green but steam distilled oil is white to pale yellowish green in appearance and scent resembles camphor. Oil yield is lower in winter compared to that in summer.

Anti-microbial properties of oil are primarily attributed to terpinen-4-ol which is well tolerated by skin. Oil also contains significant levels of 1,8 cineole (3-17%), terpinolene and γ terpinene (25%) and limonene (5%). Main constituent terpinen-4-ol varies from 25-45%. Cineole content shows much variation depending on habitat. Poor quality oil contains more cineole (10-65%) which is an irritant. It also contains α- terpinene, *p*-cymene, terpinolene, α-terpineol, α -pinene, cadinene, viridiflorene and viridiflorol.

Thyme

Scientific name: *Thymus vulgaris L.*
Family: Lamiaceae (Labiatae)
Vernacular names: *Hindi*: Banaj wain; *Kannada*: Thyme

Thyme is an ever green perennial herb, essential oil of which has a powerful fresh odour masking other unpleasant smells. This plant is extensively used as a potherb in cooking, perfumery and in liquor distillery. Thyme oil finds its major use in perfumery industry in soap and detergent work. Major constituent of oil, thymol has a powerful medicinal odour and finds more applications in flavours than in perfumes. Owing to presence of thymol, oil shows germicidal properties and is effective against a variety of pathogenic bacteria. It is employed in dental preparations, oral hygiene products, vermifuges and antigastro-intestinal products. In aromatherapy, garden thyme is regarded as one of the most important elements because of its antiseptic properties. Essence is effective in treating whooping cough and parasitic infestations. Dried leaves and floral tops constitute thyme of commerce known as *Thymi Herba* in pharmacy. Dried flowers and leaves are used to preserve linen from insects and to impart characteristic smell.

Habitat and distribution

Thyme is a native of Mediterranean region. It grows wild in almost all countries bordering Mediterranean area, Asia and Central Europe. It is extensively cultivated in Germany, France, Spain, England and various other neighbouring countries both for seasoning and for its volatile oil. Thyme production is both geographically widespread and easily undertaken. World production of thyme oil is 20-30 tonnes per annum. Spain is the largest producer followed by France, Morocco, Turkey and Mediterranean countries.

The plant

There are many species of thyme. *T. serpyllum* L. and *T. satureioides* Coss. & Bal. are some wild species. Commercial species are derived from *T. zygis* L. (white thyme) and *T. vulgaris* L. (garden thyme). Demand for garden thyme is more and hence commonly cultivated. Plant *T. vulgaris* is an evergreen perennial aromatic herb 20-30 cm high. Roots are fairly robust and stem very much branched. Sessile leaves vary in shape from elliptic to linear or diamond-shaped towards apex; young leaves are slightly woolly; flowers united in spikes at top of branches, have a bilabiate, tube-like calyx and corolla, lower lip of corolla 3-lobed; fruit consists of a smooth, dark coloured, 4-sectioned nutlet found in remains of calyx tube.

Cultivation

Plant grows the best in a warm humid climate at an elevation of 1500-4000 m from MSL. Light loamy fertile and calcareous soils are suitable. On heavy wet soils, leaves become less aromatic.

Thyme is propagated by divisions of old plant cuttings, layering or by seeds. Cuttings and layers are prepared during summer. Seeds are sown on well prepared nursery beds. Seedlings are very small and remain inconspicuous for several weeks after germination. Planting of rooted cuttings, layers or transplanting of seedlings is done during late summer at a spacing of 30-45 cm between plants and 60 cm between rows. In autumn, a light dressing of farm yard manure is given. Fertilizers are applied at 100:40:40 kg N, P_2O_5 and K_2O/ha. Top dressing of N in spring promotes formation of numerous leafy shoots. Irrigation is given when warranted. Field is to be kept weed free. Not much pests and diseases are reported in this crop. About 15 cm long shoots in early flowering stage are harvested during May-June. Lower portions of stem, together with any yellow or brown leaves are rejected.

Extraction and utilisation

Harvested herb is transported to drier immediately. Alternatively, on a smaller scale, herb can be tied in small bunches and hung on to dry in sun or in a well ventilated shed or room. Dried flowering tops are steam distilled to get thyme oil. On an average, oil recovery is 2%.

Thyme oil has following properties.

Specific gravity (25°C)	0.891
Optical rotation (25°C)	-3° 12'
Refractive index (25°C)	1.4909
Phenol content	28.0%
Solubility in 80% alcohol	7 vol.

Chemical composition of oil is as camphene + pinene 0.15%, *p*-cymene 15-50%, linalool 3-13%, linalyl acetate 0-6%, borneol 2-8%, carvacrol 0-20%, thymol 5-60%, α-thujene 0.5%, β-pinene 4.6-4.7%, myrcene 0.4-0.9%, δ-3-carene 0.1%, α-phellandrene 0.1-0.2%, limonene+1,8-cineole 35.7-44.4%, α-terpinene 0.3%, trans-linalool oxide 0.5%, cis-linalool oxide 1.0-1.1%, camphor 11.6-16.3%, β-terpineol 0.6-0.9%, α-terpineol+borneol+bornyl acetate 7.8-8.9%, α-terpinyl acetate 0.7-1.4%, geranyl acetate 0.5% and geraniol 0.1-0.2%.

Tuberose

Scientific Name: *Polianthes tuberosa* Linn.
Family: Amaryllidaceae
Vernacular names: *Sanskrit*: Rajanigandha; *Hindi*: Gulcheri; *Bengali*: Rajanigandha; *Malayalam*: Tuberose; *Telugu*: Nelasampenga; *Punjabi*: Gulshabbo

Among flowering plants which are valued much by aesthetic world for beauty and fragrance of their flowers, tuberose occupies a very special position because of its prettiness, elegance and sweet pleasant fragrance. This bulbous plant is source of tuberose oil of commerce which is very expensive and used in high grade perfumery. Fleshy, white, tubular flowers emit a strong odour and hence are cultivated on a large scale in some parts of world for extraction of highly valued natural flower oil, tuberose oil. Predominant characteristics of this crop are its lingering, delightful fragrance and excellent keeping quality. It is also cultivated for cut flowers and for preparing bouquets and garlands. Long flower spikes are excellent as cut flowers for table decoration. Individual florets are used for making garlands, floral ornaments, bouquets and button-holes. Medicinally, flowers are diuretic and emetic. Dried and powdered bulbs are used as a remedy for gonorrhea.

Habitat and distribution

Tuberose originated in Mexico or Central America. It is widely cultivated in southern France and many other tropical and subtropical areas for extraction of flower oil. In India, tuberose is cultivated on a commercial scale in West Bengal, Karnataka, Andra Pradesh and Tamil Nadu. For many years, tuberose flower oil is one of the most valuable and expensive raw materials for perfumes. Its essential oil is exported to France, Italy and other European and gulf countries.

The plant

Polianthes tuberosa has a tuberous root-stock and a mass of basal foliage. Leaves are pale-green, long, narrow and very dense. Leaves at base are 30-40 cm long and some times reddish. Long flowering stems reache up to 100 cm height. Top of stem bears pure white waxy textured racemes of blooms. Flowers are borne in pairs on apex spike and are 3-6 cm in length. Tube is long, narrow and funnel shaped, slightly bent near base. Filaments are attached to upper part of corolla. Ovary is three celled with stigmas which are ovate-fulcate. Fruit is crowned by a persistent perianth and seeds are flat.

Cultivation

There are four groups of cultivars of tuberose.

i. Single: This group is the most widely cultivated. Flower is pure white and has got a single row of corolla segments. 'Calcutta Single', 'Mexican Single', 'Rejat Rekha' and 'Suvarna Rekha' are cultivars in this group.

ii. Double: Flowers are white, tinged with pinkish red. Petals are in several whorls. 'Pearl' and 'Calcutta double' belong to this group.

iii. Semi-double: Similar to double but with only 2-3 rows of corolla segments.

iv. GVB54 Variegated: This has got variegated leaves with yellow margins.

There are single and double flowered varieties. Single flowered type is mostly cultivated for extraction of its perfume while double flowered variety usually goes to cut flower trade. Using commercially cultivated varieties "Mexican" single and "Pearl" double as parents, a high yielding F1 hybrid "Shringar" single was developed which produced 40% higher flower yield over parents containing 2.15% indole in absolute. Flowers on top of long stalk are grouped in spike-shaped clusters of 15 to 20 cm long. Flowering period begins in July reaching its maximum towards middle of August and lasts up to end of September when a secondary blooming takes place.

Although tuberose can be grown under a wide range of climatic conditions, a mild climate with an average temperature ranging from 20 to 30°C is ideal. Loam and sandy loam soils having a pH range of 6.5 to 7.5 with good aeration and drainage are the best suited for cultivation. Spindle shaped disease free bulbs having a diameter of 1.5-3.0 cm is used for planting. Mother bulbs are the best for planting as they flower early. Finger or side bulbs take 2-3 years to come to flower. Single type has maximum fragrance and is popular among growers for production of essential oil. The best time of planting is from May to July. Land is ploughed 2-3 times and soil is brought to fine tilth. Well-rotten FYM at 20-30 tonnes/ha is applied and mixed well. Furrows are opened 25-30 cm apart and bulbs are planted at 25 cm spacing in furrows. About 1.25 lakhs/ha (800-900 kg) of bulbs are required for planting. A fertilizer dose of 100:200:200 kg N, P_2O_5 and K_2O/ha is generally recommended. Half dose is applied basally and other half as topdressing when flower spikes start appearing. Application of GA_3 (1000 ppm) induces early flowering, increased flowerstalk production and improves quality of flowers. Weekly irrigation and regular weeding are required for the best yield. In order to keep plots free of weeds and to avoid exposure of bulbs, plots are weeded and earthed up once a month.

Flowering season is between June and October. Flowers will be ready for harvest in 3-3.5 month's time. They are harvested by cutting fully opened spikes from base during cool hours of day either in morning or evening. From one planting, 2-3 ratoons can be taken for which flower stalks of main crop are headed back and plot is manured and irrigated. Average yield comes to 5-10 tonnes/ha for planted crop, 9-12 tonnes/ha for first ratoon and 4-6 tonnes/ha each for subsequent ratoons. Wider spacing showed marked improvement in plant growth and flowering of tuberose. Wider spacing of 45x30 cm resulted in significantly higher spike diameter, leaf area and number of bulbs in main and ratoon crop. Instances of a ratoon crop producing more number of spikes/unit areas in comparison to main crop are recorded. Flowers can be kept in poly ethylene bags for maintaining freshness and retaining white colour for longer time.

Pest and disease management

Pests

Grass hoppers and caterpillars feed on leaf lamina. Thrip sucking sap from inflorescences is another pest of importance. These can be controlled by spraying mercaptothion 50EC (2 ml/l) or carbaryl 50WP (4 g/l). Red spider mites damaging leaves are effectively managed by spraying of dicofol 18.5EC (3ml/l) or wettable sulphur (1g/l). Nematode *Meloidogyne incognita* and greecy streak nematode *Aphelencoides besseyi* cause stunted growth of plants. Soil application of carbofuran 3G @ 20 kg/ha reduces damage.

Diseases

Root and basal rot caused by different species of fungi is the major disease in tuberose. If proper control measures are not taken, plants get wilted and die. As a prophylactic measure, planting materials should be dipped in captafol (0.3%) for one hour before planting. If the disease prevail in main field, soil drenching with hexaconzole (0.2%) is effective

Leaf spot caused by fungus *Alternaria polyantha* and blight caused by *Botrytis elliptica* are also major diseases.

Extraction and utilisation

Tuberose is one of those plants, flowers of which continue to develop their natural fragrance for some time after they have been harvested. Flower oil is extracted by enfleurage and solvent extraction with petroleum ether. Distillation cannot be employed as steam or water distillation of tuberose flowers directly gives only a very little yield of oil and it is of very poor odour. Freshly picked flowers, before they open are enfleuraged. About 150 kg of flowers yield 1 kg of absolute of enfleurage which contains 11-15% of steam volatile oil. Extraction of tuberose flowers with petroleum ether yields 0.08-0.14% of concrete. Concrete contains 3-6% of a steam volatile oil.

Absolute of enfleurage is a brown, semi-solid and alcohol soluble liquid possessing a characteristic odour of tuberose flowers with a fatty by-note. It has specific gravity (15°C) 1.009-1.035, refractive index (20°C) 1.535-1.574, optical rotation -2° 30', acid value 32.7 and ester value 243-280.

Concrete of tuberose is a light to dark brown waxy mass which is only partially soluble in high-proof alcohol. It has congealing point 49-57°C, melting point 57°C, specific gravity (60°C) 0.8951, refractive index (60°C) 1.4601, acid value 52-56, ester value 63-76 and saponification value 117-119.

Chemical constituents of tuberose flower oil include geraniol, nerol, farnesol, benzyl alcohol, methyl benzoate, benzyl benzoate, methyl salicylate, methyl anthranilate, eugenol and butyric acid.

Ylang-Ylang

Scientific name: *Cananga odorata* (DC) Hook. F & Thoms
Family: Anonaceae
Vernacular names: *Malayalam*: Pachachempakam; *Tamil*: Manoranjitham; *Kannada*: Apoorva sakpaka; *Telugu*: Apoorva sampakamu

Two types of oil are produced from flowers of *C. ordorata*; cananga oil from forma macrophylla and ylang-ylang oil from forma genuine. Oil of ylang- ylang is highly appreciated in perfumery because of its delightfully sweet and strong odour. Main use of cananga oil is in soaps where its tenacity is valuable, toiletries especially those for men and less expensive perfumeries and similar products. Ylang-ylang oil is normally available in four grades extra, first, second and third. Extra grade is extensively used in high quality perfumes where its powerful odour enables inclusion in very small amounts. Third grade oil is extensively used in less expensive soaps, toiletries and fragrances.

On steam distillation, fresh flowers of the semi wild and cultivated trees yield an essential oil with a strong and pleasant odour, greatly appreciated and widely used in perfumery.

Habitat and distribution

Original home of Cananga is probably south east Asia, although it is now naturalized in Burma, Malaysia, Indonesia, Papua New Guinea and other Pacific islands and Philippines. Cananga is introduced into tropical countries in Africa, Asia, the Caribbean, the Americas and the then French colonies of Indian Ocean, specifically as an essential oil plant. Commercial cultivation and oil production are now concentrated in Indonesia and Madagascar with small amounts of oil from Reunion Comoro and Philippines.

The plant

Genus *Cananga* has two species; C. *odorata*, the source of cananga and ylang-ylang oil and *C. latifolia* Gins Gay Nep., producing an oil of commercial importance. Name of the tree indicates drooping nature of flowers. Leaves are numerous, thin, alternate, with a sharply pointed apex, base rounded, margin entire, 15-20x4-7 cm on light green petioles. Flowers are numerous, large, yellow-green, strongly scented and borne on slender light green stalks in axillary clusters. Individual flowers bear a narrow pendulous structure composed of three ovate-hairy sepals and six almost lanceolate petals with slightly recurved tips, 4-8 cm in length, arranged in two rows. Flower center is composed of separate carpels arranged in two series, each carpel containing a single ovule developing into separate fruit.

Cultivation

The tree requires a moist tropical climate. It grows well in rich volcanic soils or fertile sandy loams. In Java, cananga tree occurs largely wild and is planted occasionally. Propagation is largely by seeds. Seeds are sown in a seed bed in March. After 3-4 months, seedlings are planted in a polythene bag and watered frequently. Four to six months old seedlings are planted in main field at a spacing of 6x6 m. Plants are to be shaded in summer. Propagation usually takes place spontaneously by means of fruit (seed) falling off the trees or by birds eating the fruit and dropping the seed elsewhere. Natives may find a spontaneously growing young plant and they transplant it to garden without giving it any special care. Rarely and only when prices of cananga oil are very attractive, natives plant new trees grown from one year old seeds. Fresh seeds do not germinate well. During rainy season, seed is planted into holes and fertilized with cow dung. Young plants develop more or less rapidly, depending upon climate. At low altitudes, trees require only a few years to give an economic flower harvest, but at higher and cooler altitudes (500 m above sea level) they grow much more slowly and require seven to ten years to produce flowers in quantities sufficient for harvesting.

After two years first bunch of flowers appear. By beginning of third year, the trees would have achieved a height of 2-3 m and then topped. This removal of epical dominance encourages side shoot formation into lateral branches. Periodic pruning has to be undertaken to prevent growth of the tree beyond 2-3 m height. Commercial production of flowers starts from fourth year of planting. Economic life span goes up to 25 years for a well managed plantation. No serious pest or disease is noticed in the plant.

Cananga tree possesses a deep taproot; if the roots encounter hard subsoil, crown of the tree will not develop well. Trees of more than twenty years old reach a height of about 30 m. Trees flower from second year, when a small harvest may be taken and flower profusely from fourth or fifth year. A seven year old tree bears from 30 to 100 kg of flowers per year. A fully grown tree produces up to 300 kg of flowers depending on climatic conditions.

Though the tree flowers throughout year, three main flowering seasons are seen for commercial production of oil. Principal season is immediately after rainy season, a moderate harvest season during dry season and rainy season. Flowers contain more essential oil during night, particularly just before day break and hence harvesting is done during early morning hours. Flowers are manually harvested and only fully developed yellow flowers are gathered. Great care is to be taken not to crush the flowers during harvest.

Flowers have little fragrance initially. Fine white hairs covering green petals disappear with maturity. After 14-21 days, petals become pale green, then yellow and their scent becomes progressively stronger. When two small reddish spots appear on base of petals, the flowers are ready for picking. This should be as early in morning as possible, when oil content is highest and normally ceases by 10 am as in strong sunlight, oil content rapidly diminishes. Composition and characteristics of oil change as flowers develop, with light oxygenated compounds the highest at maturity; thus inclusion of immature flower is detrimental to oil quality.

Extraction and utilisation

Flowers should be immediately distilled for oil since flowers tend to fade and ferment on keeping. Fractional distillation of oil is carried out normally. First fraction of the oil contains most of the aromatic constituents of oil such as esters and ethers, whereas later fractions consist chiefly of sesquiterpenes which have little odour value. In retort, water is taken and heated to about 70°C and then flowers are quickly put into retort. Distillation is started slowly by carefully injecting live steam through perforated steam coil. The volatile oil distils out easily and first fraction which consists chiefly of esters and ethers are collected in oil separator. After a while, direct steam is shut off, distillation being continued by heating with indirect steam. Fractions are cut according to specific gravity into extra (0.955), first (0.942), second (0.922) and third fractions (0.910-0.912). Distillation is continued for a period of about 18-22 hours. Yield of oil ranges from 2-2.5%.

Distillation is done frequently in small and direct-fired water stills. Heating water to near boiling prior to adding flowers results in higher quality oil. Still bodies should preferably be under charged with flowers, since this reduces spot fermentation or over heating either being detrimental to oil quality. Distillation by any method must be carefully controlled since oil factions in distillate have different characteristics and market value.

Weight of flowers to produce 1 kg crude oil varies from 350-700 kg depending mainly on time of year and flower maturity. In a modern distillery, a total yield of 2.0-2.5% is usual, about 25-30% being the first two grades. Flowers can be solvent extracted with petroleum either or benzene to give a yield up to 1.2% and resultant concentrate has an odour virtually identical to natural flower. Air-drying of Philippine flowers for 1 to 7 days before distilling changes specific gravity (20°C) from 0.03-0.96 to 0.913-0.914 and ester number from 120-190 to 145-150. In general, specific gravity, total esters and saponification number decrease, while optical rotation significantly increases with decrease in grade. Lower grades are virtually insoluble in 90% alcohol.

Main characteristics of the concentrate are specific gravity (20°C) 1.020-1.035; optical rotation -5.5° to -6.5°; refractive index (20°C) 1.50-1.52; ester number 200-250; but quoted data vary widely for no apparent reason.

References

Anzaldo, F.E. 1980. Essential oil production in the Philippines. NSDB Tech. I. 5: 67-71

Arya, P.S. 1983. Effect of harvesting dates on the quality of Hops (*Humulus lupulis* L.). Indian Perfumer. 27: 12-13.

Ashraf, M. and Bhatty, M.K.1975. Studies on the essential oils of the Pakistan species of the family Umbelliferae, 1. *Trachyspermum ammi* (L.) Spargue (Ajowan) seed oil. Pak. J. Scient. Ind. Res., 18: 232-235

Atal, C.K. and Kapur, B.M. 1982. *Cultivation and Utilization of Aromatic Plants*. Regional Research Laboratory, Jammu-Tawi, 815p

Bahl, J.R., Bansal, R.P., Fatima, A., Garg, S.N., Singh, V. and Kumar, S. 2000. Growth and essential oil characteristics of *Lippia alba* in sub tropical environment. J. Med. and Arom. Plant Sci., 22: 739-742

Bahl, J.R., Garg, S.N., Singh, S.C., Bansal, R.P., Naqvi, A.A. and Kumar, S. 2000. Compositions of linalool rich essential oil from *Lippia alba* grown in Indian plains. Flav. Frag. J. 15: 199-200

Bakshi, S.K., Bhan, M.K. and Kitchlu, S. 1991. Studies on the stage of harvesting in five exotic varieties of Hops (*Humulus lupulus* L.). Indian Perfumer 35: 218-220.

Bassett, I.B., Panaowitz, D.L. and Barnetson, R.S. 1990. A comparative study of tea tree oil versus benzoil peroxide in the treatment of acne. Med. J. Aust., 153: 485-458

Bhati, A. 1953. Essential oil from berries of *J. communis* Linn. J. Indian Inst. Sci., 35: 43-46

Biddington, N.L. and Thomas, T.H. 1979. Residual effects of high temperature pretreatments on the germination of celery seed. Physiologica Planta., 47: 211-214

Birsat, M., Sarma, S. and Maheswari, M.L. 1997. Composition of essential oil from the flowers of Kewda. Indian Perfumer, 41: 69-72

Boelens, M.H. 1985. The essential oil from *Rosemarinus officinalis* L. Perfumer Flavorist, 10: 21-38

Carson, C.F. and Riley, T.V. 1995. Anti-microbial activity of the major components of essential oil of *Melaleuca alternifolia*. J. Appl. Bacteriol., 78: 264-269

Chandra, V. and Srivastava, H.P. 1977. Package of practices for cultivation of *Jasminum grandiflorum* L. Indian Perfumer, 21: 1-3

Chao, S.C. and Young, D.G. 2000. Screening for inhibitory activity of essential oils on selected bacterial, fungi and viruses. J. Essen. Oil Res., 12: 639-649

Charles, D.J., Simon, J.E. and Singh, N.K. 1992. The essential oil of *Alpinia galanga* willd. J. Essen. Oil Res., 4: 81-82

Chauhan, H.S., Patra, N.K., Mengi, N., Singh, H.P. and Singh, A.K. 2001. Yield maximization in *Rosa damascena* var. Noorjahan. Indian Perfumer, 45: 217-221

Chopra, R.N., Nayar, S.L. and Chopra, I.C. 1956. *Glossary of Indian Medicinal Plants.* Council of Scientific and Industrial Research, New Delhi, 330p

Christopher, T.A., Hoffman, A. and Allen, O. 1986. *Magic and Medicine of Plants.* The Readers Digest Association, Inc. Pleasantville, New York, Montreal, 142p.

CSIR, 1953. *The Wealth of India, Industrial Products Part III D-E.* Council of Scientific and Industrial Research, New Delhi, 250p

CSIR, 1966. *The Wealth of India. Raw Materials Vol. VIII.* Council of Scientific and Industrial Research, New Delhi

Damle, AV. and Tipnis, M.P. 1980. Chemical and antimicrobial investigation of callus culture of *Trachyspermum ammi* Spargue (Ajowan). Ind. J. Pharm. Sci., 42: 86-88

Davies, N.W. 1990. Gas chromatographic retention indices of monoterpenes and sesquiterpenes on methyl silicon and carbowax 20M phases. J. Chromato. 503: 1-24

Duke, J.A. and Reed C.F. 1978. Caraway- an economic plant. J. Crude Drug Research. 16: 116-118.

Dung, N.X., Khien, P.V., Nhuan, D.D., Hoi, T.M., Ban, N.K., Leclereq, P.A., Muselli, A. and Cassanova, J. 1999. Composition of the seed oil of *Hibiscus abelmoschus* L. (Malvaceae) growing in Vietnam. J. Essential Oil Res., 11: 447-452

Dwivedi, S. K., Kishore, N. and Dwivedi, S. K. 1990. Fungitoxicity of some essential oils against *Macrophomina phaseolina.* Indian Perfumer, 34:20-21

Elakovich, S.D., Ogantimein, B.O. 1987. The essential oil of *Lippia odoensis* leaves and flower. J. Nat. Prod., 50:503-506

Farooqi, A. A., Khan, M. M. and Vasundhara, M. 1999. *Production technology of Medicinal and Aromatic Crops.* Natural Remedies Pvt. Ltd., Bangalore, 176p

Farooqi, A. A. and Sreeramu, B.S. 2001. *Cultivation of Medicinal and Aromatic Crops.* Universities Press (India) Ltd., Hyderabad, 518p

Fehr, D. 1974. Essential oil from leaves of celery (*Apium graveolens*). Pharmazie, 29:349

Frighetto, N., Oleviera, J.G., Siani, A.C. and Chagas, K.C.D. 1998. *Lippia alba* Mill. NE Br (Verbenaceae) as a source of linalool. J. Essential Oil Res., 10: 578-580

Fun, C.E. and Svender, A.B. 1990. The essential oil of *Lippia alba* (Mill) NE. Br. J. Essential Oil Res., 2: 265

Garg, S.C. 2004. A Review on biological activity of tea tree oil. Indian Perfumer, 48: 195-197

Garg, S.N., Gupta, D., Sadhidhara, K.V. and Mehta, V.K. 2003. *Alpinia galanga* L.- Composition of essential oil from flowers. Indian Perfumer, 47: 147-149

Grieve, M. 1992. *A Modern Herbal.* Tiger Books International, London, 912p

Guenther, E. 1949. *The Essential Oils.* Vol. III. Van Nostrand Co. Inc., New York, 777p

Guenther, E. 1950. *The Essential Oils.* Vol. IV. Van Nostrand Co. Inc., New York, 752p

Guenther, E. 1952. *The Essential Oils.* Vol. V. Van Nostrand Co. Inc., New York, 507p

Guenther, E. 1952. *The Essential Oils.* Vol. VI. Van Nostrand Co. Inc., New York, 481p

Gupta, G.N., Chandra, G. and Dhingra, D.R. 1955. Otto of rose. Indian Oil Soap J., 21:61-68

Gupta, R. 1965. Introduction of Hops cultivation in Himachal Pradesh. Indian Perfumer, 9:81-85.

Handa, K.L., Chopra I.C. and Sobti, S.N. 1957. Aromatic plant resources of Jammu & Kashmir. J. Scientific Res., 16:1-28

Husain, A., Virmani, O.P., Sharma, A., Kumar, A. and Misra, L.N. 1988. *Major Essential Oil Bearing Plants of India.* Central Institute of Medicinal and Aromatic Plants, Lucknow, 237p

ISI, 1965. IS:3146-*Oil of celery seed.* Indian Standard Institute, New Delhi, India

ISI, 1965. IS:3147-*Oil of dill.* Indian Standard Institute, New Delhi India.

ISI, 1965. IS:3227-*Oil of bois de rose.* Indian Standard Institute, New Delhi, India

Joshi, S.G. 2000. *Medicinal Plants.* Oxford Publishing Co.Pvt. Ltd., New Delhi, 491p

Joy, P. P, Thomas J., Mathew, S. and Skaria, B. P. 2001. Medicinal Plants. In: *Tropical Horticulture Vol. 2.* Bose, T.K., Kabir, J., Das, P. and Joy, P.P. (eds.), Naya Prokash, Calcutta, India, pp 449-632

Joy, P. P., Thomas, J., Mathew, S., Jose, G. and Joseph, J. 2001. Aromatic plants. In: *Tropical Horticulture Vol. 2.* Bose, T.K., Kabir, J., Das, P. and Joy, P.P. (eds.), Naya Prokash, Calcutta, India, pp 633-733

Joy, P.P., Thomas, J., Mathew, S. and Skaria, B.P. 2001. Influence of harvest duration on the yields of rhizome, root, shoot and their oils in *Alpinia galanga.* J. Med. and Arom. Plant Sci., 23: 341-343

Junimann, M. 1999. *Enchanting Scents.* Sri Satguru Publication, Delhi, 123p

Kahol, A.P. 1985. Techno-economic process for production of rose oil. Indian Perfumer, 29: 37-42

KAU, 2003. Package of Practices Recommendations 'Crops'-2003. Directorate of Extension, Kerala Agricultural University, Thrissur. 278 p

Kaul, V.K., Singh V. and Singh, B. 2000. Damask rose and marigold - prospective industrial crops . J. Med. and Arom. Plants. Sci., 22: 313-318

Kumar, A., Sharma, A. and Virmani, O.P. 1985. Cultivation and utilization of rose geranium: A review. Cur. Res. Med. Arom. Plants, 7: 137-147

Kumar, S., Bahl, J.R., Bansal, R.P., Garg, S.N., Naqvi, A.A., Khanuja, S.P.S., Shasany, A.K., Singh, V. and Singh, S.C. 1999. Linalool rich *Lippia alba* var. Kavach. J. Med. and Arom. Plant Sci., 21: 1094-1095

Kumar, S., Sharma, P. K., Yadav, G. L. and Sharma B.L. 2001. Influence of sowing time on the performance of Dill (*Anetheum graveolens*) under rainfed conditions. J. Med. Arom. Plant Sci., 23: 34-35

Lawrence, B.M. 1980. New trends in essential oils. Perfumer and Flavorist, 5: 6-16

Mahindru, S.N. 1992. *Indian Plant Perfumes.* Metropolitan, New Delhi, 263p

Mallavarapu, G.R., Bahl, J.R., Bansat, R.P., Garg, S.N, Ramesh, S. and Kumar, S. 2000. Essential oil of *Lippia alba,* a rich source of linalool. J. Med. Arom. Plant Sci., 22: 765-767

Mishra, A.C., Negi, K.S., Sunaja, P. and Maheswari, M.L. 2004. Performance of Marjoram (*Marjorana hortensis*) in UP. Indian Perfumer, 48: 41-45

Misra, B.C., Rao Y.R., Reeta M., Panighrahi M.R. and Moharana, C. 1998. Economic importance of Kewda *(Pandanus odoratimuss*) in Ganjan district. Indian Perfumer, 42: 128-130

Mwangi, J.W., Mensah, A.I., Muriuki, G., Munavu, R., Lwande, W. and Hassanali, A. 1992. Essential oil of *Lippia* species in Keny N Maize weevil (*Sitophilewa Zeamais*) repellency and larvicidal activity. J. Pharmacog. 30: 9

Nagaraja, G.S., Gowda, N. and Farooqi, A.A. 1998. Studies on tuberose (*Polianthes tuberosa* L.) cv. single for higher yield and shelf life. Indian Perfumer, 42: 158-164

Nath, S.C., Hararika, A.K. and Bordoloi, D.N. 1988. Citral in *Litsea cubeba* Pers. from Arunachal Pradesh. Environmental Awareness, 11: 19-20

Nerle, S.K. and Torne, S.G. 1984. Studies in *Kaempferia galanga* L. Indian Drugs, 21: 236-237

Nigam, M.C. and Misra, L.N. 1980. Chemistry of jasmin oil-A review. Curr. Res. Med. Arom. Plants, 2: 37-44

Pal, B.P. 1972. *The Rose in India.* Indian Council of Agricultural Research, New Delhi

Pandey, G. 1992. *Medicinal Flowers.* Sri Satguru Publications, Delhi, 209p

Prasad, A., Krishna A., Grag, S.N., Bhasney, S. and Singh, D.V. 2003. Growth and yield and cat ion accumulation in *Tagetes* (Marigold) species under solicit soils. J. Med. Arom. Plant Sci., 25: 408-413

Prochazka, F. 1981. Caraway cultivation in largescale agricultural production. *Nasc Liecive* Rastiny, 18: 82-86

Raina, V.K., Kumar, A., Tandon, S., Ahmad, J. and Kahol, A.P. 2005. Composition of juniper berry oil of commerce. Indian Perfumer, 49: 329-332

Rajagopalan, A. and Gopalakrishnan, P.K. 1985. Growth, yield and quality of *Kaempferia galanga* as influenced by planting time and types of seed material. Agric. Res. J. Kerala, 23: 83-89

Rajagopalan, A., Viswanathan, T.V. and Gopalakrishnan, P.K. 1989. Phytochemical analysis and nutrient uptake studies on *Kaempferia galanga* L. South Indian Hort., 37: 34-38

Ram, V.A. 1997. *Herbal Indian Perfumes and Cosmetics*. Sri Satguru publications, 165p

Ranade, G. S. 1980. The rose fragrance. Indian Perfumer, 24: 49-56

Ranade, G. S. 1998. Anethi oil (Indian Dill Seed). FAFAI, 20: 61

Ranade, G.S. 2002. Essential oil profile (juniper berry oil). FAFAI, 4: 51

Rao, Y.R. 2000. 'Kewda *(Pandanus fascicularis)*, an economically important aromatic shrub in Ganjan district, Orissa, India., Regional Research Laboratory, Bhuvaneswar. J. Med. Arom. Plant Sci., 22: 377-395

Rishi, A.K., Dhar, P.L., Nasir A. and Bhan, M.K. 1985. Intercropping of *Dioscorea deltoidea* with Hops (*Humulus lupulus* L.). Herba Hung. 24: 69-75.

Rout, P.K., Mohapatra, D.K., Shahoo, S. and Nanda, B. 2004. Seasonal variation in composition of elite Kewda genotype essential oil by GC -MS analysis. J. Med. Arom. Plant Sci., 36: 731-733

Sarin, Y.K. 1977. Skimmia leaf: A potent source of essential oil from hill area of Uttar Pradesh. In: *Advances in Essential Oil Industry*. Kapoor, L. D. and Ram Krishnan (eds.), Today and Tomorrow's Printers and Publishers, New Delhi

Sen, 1996. Occurrence of *Alternaria* leaf spot and flower blight on marigold (*Tagetes erecta*) in Himachal Pradesh. Indian Perfumer, 40: 27

Sharma, A., Kumar, A. and Virmani, O. P. 1985. A review on clarysage (Salvia sclarea L.). Cur. Res. Med. Arom. Plants, 7: 39-48.

Sharma, M.L., Singh, A. and Taneja, B.R. 1980. *Extraction of Jasmine flower oil-Extension Bulletin No. 7*. National Botanical Research Institute, Lucknow, India

Sharma, P.C., Yelne, M.B. and Dennis, T.J. 2001. *Database on Medicinal Plants Used in Ayurveda. Vol. 2*. Central Council for Research in Ayurveda & Siddha, New Delhi, 590p

Sharma, R.N. 1991.Contribution of Ayurveda towards management of bronchial asthma. Sachitra Ayurveda, 44: 349-352

Singh, A., Singh, M. 2004. Weed management in essential oil yielding Damask rose (*Rosa damascenea)*. J. Med. Arom. Plant Sci., 26: 71-726

Singh, A.K. 2002. Weed biomass, Rose plant growth and flowering as influenced by weed control measures. Indian Perfumer, 463: 235-239

Singh, G., Singh, O.P., Maurya, S., Marimuthu, P., Lampasona, M.P.de and Catalan, C. 2005. Composition of needle oil of *Juniperus communis* L. from northern India. Indian Perfumer, 49: 163-167

Singh, M. and Ramesh, S. 2000. Effect of irrigation and nitrogen on herbage, oil yield and water use efficiency in rosemary grown under semi-arid tropical conditions. J. Med. Arom. Plant Sci., 22: 659-662

Singh, T.A., Ram, P. and Husain, A. 1979. *Hops and its Cultivation in India-Farm Bulletin No. 13*. Central Institute of Medicinal and Aromatic Plants, Lucknow, 31p

Singh,G., Rao, G.P., Kapoor, I.P.S. and Singh, O.P. 2000. Chemical constituents and antifungal activity of *Lippia alba* Mill. leaf essential oil. J. Med. Arom. Plant Sci., 22: 701-703

Sorm, F. 1949. Composition of oil of Hops. Collczech. Chem.Commun., 14: 693-723.

Srinivas, M. and Murthy, N. 1997. High yielding tuberose (*Polianthes tuberosa* L.) hybrid- "Shringar" for concrete. Indian Perfumer, 41: 157-161

Srivastava, S.K. and Nigam, M.C. 1979. Chemistry of rose oil: A review. Curr. Res. Med. Arom. Plants, 1:147-155

Stashenko, E.E. 1995. A study in the compositional variation of the essential oil of Ylang-Ylang (*C. odorata* forma *germina*) during flower development. J. High Resol.Chromatog. 18:101-104

Subbiah, B.V. and Asija, A.L. 1956. A rapid procedure for estimation of available nitrogen in soil. Curr. Sci., 25:258

Swaminathan, K.R.., Muthuswamy, S. and Rao, V.N.M. 1979. Pilot plant for extraction of jasmine essential oil from *Jasminum grandiflorum*. Indian Hort, 24: 20-21

Tandon, S., Kahol, A.P., Kumar A. and Ahamed, J. 2005. Rose concrete and absolute from *Rosa damascena* var. Noorjahan . Indian Perfumer, 49: 77-81

Taskinen, J. and Nykanen, L. 1976. Volatile constituents of an alcoholic extract of Juniper berry. International Flav. Food Addit., 7: 228-229

Thakur, R.S., Puri, H.S. and Husain, A. 1989. *Major Medicinal Plants of India*. Central Institute of Medicinal and Aromatic Plants, Lucknow, India, 585p

Thomas, J., Joy, P.P. and Mathew, S. 1997. Cultivation and utilisation of *Kaempferia galanga* Linn. In: *Supplement to Cultivation and Utilisation of Aromatic Plants.* Handa, S. S. and Kaul, M. K. (eds.), Regional Research Laboratory, Jammu-Tawi, India, pp.299-305

Thomas, J., Joy, P.P., Mathew, S., Skaria, B.P., Duethi, P.P. and Joseph, T.S. 2000. *Agronomic Practices for Aromatic and Medicinal Plants.* Directorate of Arecanut and Spices Developmet, Calicut, 124p

Verma, N., Tripathi, A.K., Prajapathi, V., Bahal, J.R., Bansal, R.P., Khanuja, S.P.S. and Kumar, S. 2000. Toxicity of essential oil from *Lippia alba* towards stored grain insects. J. Med. Arom. Plant Sci., 22:117-119

Weaver, D.K., Dunkel, F.W., Ntezurubanza, L., Jackson, L.L. and Stock, D.T. 1991. The efficacy of linalool, a major component of freshly milled *Ocimum canum* (sims)

chamiacea, for protection against stored produce Coleoptera. J. Stored Prod. Res. 27: 213-220

Weiss, E.A. 1997. *Essential Oil Crops*, CAB International, New York, 600p

Wilson, C.W. 1969. Terpene and sequiterpene hydrocarbons in the essential oil of fresh celery. J. Food Sci., 34: 521-523

Wony, K.C., Ong, K.S. and Lim, C.L, 1992. Composition of the essential oil of rhizomes of *Kaempferia galanga* L. Flavour and Fragrance J., 7: 263-266

Yadava, R.N. and Saini, Y.K. 1991. Gas chromatographic examinations of leaf oil of *Marjorana hortensis*. Indian Perfumer, 35: 102-103

Ajowan (Page - 149)

Ambrette (Page - 151)

Caraway (Page - 155)

Celery (Page - 158)

Champak (Page - 161)

Clarysage (Page - 164)

Dill (Page - 167)

Galangal (Page - 171)

Greater Galangal (Page - 174)

Hops (Page - 177)

Jasmine (Page - 181)

Juniper (Page - 186)

Kewda (Page - 189)

Lippia (Page - 192)

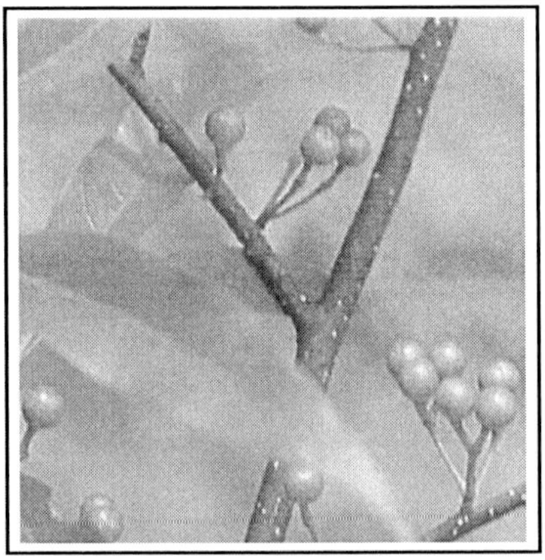

Litsea (Page - 194)

Marigolds (Page - 197)

Rose (Page - 201)

Rosemary (Page - 207)

Skimmia (Page - 210)

Sweet Flag (Page - 213)

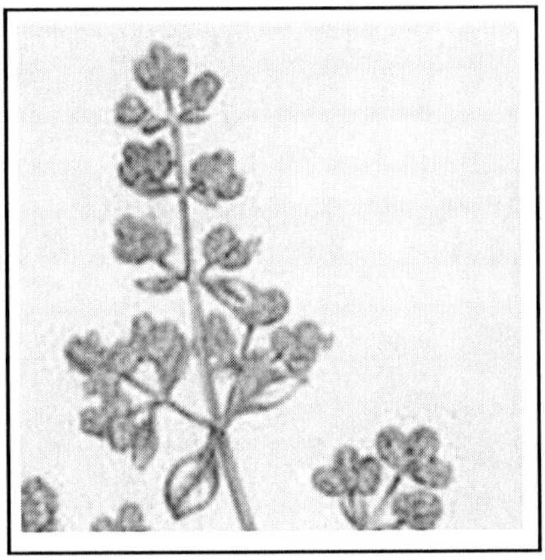

Sweet Marjoram (Page - 216)

Tea Tree (Page - 219)

Thyme (Page - 223)

Tuberose (Page - 225)

Ylang-Ylang (Page - 229)

Index
(*English Name*)

D

E

F

G

H

I

J

K

Index
(*Scientific Name*)

B

C

D

E

F

G

H

I

J

K

L

M

N

O

P

Q

R

S

T

U

V

Index
(*General Name*)

D

E

M

N

O

P

Q

R

S